Alice Wallace
897-3681

Instructor's Guide
with
Objectives, Answer Key, and Test Masters

Applied Mathematics

by

R. Jesse Phagan

South Holland, Illinois
THE GOODHEART-WILLCOX COMPANY, INC.
Publishers

TABLE OF CONTENTS

International Standard Book Number 0-87006-824-5

1234567890-92-987654321

TEACHING APPLIED MATHEMATICS

APPLIED MATHEMATICS, published by Goodheart-Willcox, is intended for students with a wide range of math skills. Through the use of problems relating to a variety of everyday problems, students are able to relate math to something useful.

Math students frequently ask, "Where will I ever use this stuff?" or "How will this math help me in my everyday life?" This book answers such questions. Furthermore, in learning these basic math skills and understanding their significance in the "real world," a student gains the confidence to continue to solve problems in real-life situations.

EXPERIENCE REAL-LIFE SITUATIONS

All aspects of education should be tied together to aid the student in a more complete learning experience. This trend is being seen nationwide with Technology Education and many of the vocational courses. The more a student can relate to the learning task at hand, the more desire to work hard at learning and remembering.

As each section is studied, the problems contained in the textbook (and optional workbook) should be practiced and reviewed. As an additional learning assignment, because the follow-through of a situation from conception to solution is a critical learning experience, ask the student to make up word problems of his/her own. The following is a suggested format:

- Content should be a situation from the student's own life experience.
- Describe the situation with words and drawings.
- Use mathematics from the textbook section being studied.
- Show the work necessary to solve the problem.

USING THE STUDENT WORKBOOK

The student workbook contains problems ranging in degrees of difficulty. Contained in each chapter are problems to help build confidence, to practice math concepts, to develop skills, and to challenge all levels of ability.

The instructor should be selective in which problems are assigned for homework. Also, it is very important for the instructor to review the homework assignments in class. Frequently, a student will have a problem completed but not fully grasp how and why the steps are performed. When the instructor reviews the problem, the student can gain new insight and will retain a higher skill level.

The student should write in the workbook and keep all of the work for reference. In this way, it is much easier to study for tests and exams. More importantly, when a real-life situation develops, the student will have the resources to solve the problem. Many school districts do not allow a student to keep a textbook, but the student is allowed to keep a write-in workbook. What better way is there to refresh your memory than by reviewing your own work? This will also help develop a better appreciation for the role math plays in a person's life.

DEVELOPING NEW TESTS

Contained within this Instructor's Guide are Test Masters for each chapter of the textbook. These tests can be changed easily by simply changing the numbers in the problem. In this way, variations of the same test can be made for security purposes while maintaining the same degree of difficulty.

SHOW THE WORK! By a student being forced to show the work when solving a problem, it prevents cheating. It also insures that the student actually knows how to work the problem and did not just guess at the solution.

OVERVIEW OF THE TEXTBOOK

The major emphasis of this math book is clearly on applications. The progression through the book allows the student to continually develop solid math skills. Depending on the nature of the course where this book is used, some chapters may be optional. However, for the most part, the chapters included here are skills needed for most every vocational and technical field.

PART I — BASIC MATH (Chapters 1-4)

There is a natural tendency for the student and teacher to want to skip the earlier chapters. The first four chapters provide the core material for understanding mathematics at all levels. Although these concepts are taught in earlier grades, students often lack the necessary skills. These chapters are of the utmost importance to developing good math skills.

PART II—APPLIED BASIC MATH (Chapters 5-9)

The concepts learned in Percentage (Chapter 4) are applied in Personal Finances (Chapter 5) and Graphs

(Chapter 6). These chapters help to learn to deal with money and other factors affecting life in general. In these chapters, students relate quite easily to how math affects them.

Chapters 7, 8, and 9 deal primarily with measurement. These chapters are very important for some vocational and technical areas and less important for others.

For real-life activities, the students should have no trouble relating this entire section to situations in their own lives. The instructor should plan extra time for these chapters to allow the students to "digest" the skills learned.

PART III—BASIC AND APPLIED ALGEBRA (Chapters 10-14)

It is more difficult to find real-life situations that need to be solved by algebra. These five chapters bring algebra to the student in a way that will keep the interest. It is a good opportunity to fine tune the math skills learned in earlier chapters while building confidence in solving problems. This should finally answer the question "How does this apply to my life?"

PART IV—ADVANCED APPLIED MATH (Chapters 15-18)

As the name suggests, this section applies the skills learned in earlier chapters to more advanced subject areas. Two areas needing strong math skills are electricity/electronics and construction. It is these areas that are stressed, although other areas of study will benefit from the advanced skills training.

Excellence in education begins with a solid foundation in basic skills.

OBJECTIVES AND ANSWER KEY

Chapter 1
SOLVING WORD PROBLEMS

Objectives

After studying this chapter, the student will be able to:

- Identify operations needed to solve word problems.
- Set up word problems and perform the required arithmetic.
- By checking, state whether or not answers to word problems are correct.

ANSWERS TO TEXT, page 16

1. 1260 *lin. ft.* in 90 min.

   $$\begin{array}{r} 14 \\ \times\ 90 \\ \hline 1260\ ft. \end{array}$$

2. 14 *A* total current

   $$\begin{array}{r} 8 \\ 2 \\ 3 \\ +\ 1 \\ \hline 14\ A \end{array}$$

3. 4520 *lb.* for the load

   $$\begin{array}{r} 6870 \\ -\ 2350 \\ \hline 4520\ lb. \end{array}$$

4. 3 *ft.* window opening

   $$\begin{array}{r} 5 \\ 3 \\ 4 \\ +\ 2 \\ \hline 14\ ft. \end{array}$$ for other dimensions

   $$\begin{array}{r} 17 \\ -\ 14 \\ \hline 3\ ft. \end{array}$$ remaining

5. *$2 per can*

   $24\overline{)48}$ = $2 each

6. $85 total cost

   $$\begin{array}{r} 18 \\ \times\ 3 \\ \hline \$54 \end{array}$$ pipe $$\begin{array}{r} 4 \\ \times\ 1 \\ \hline \$4 \end{array}$$ elbows

   $$\begin{array}{r} 2 \\ \times\ 6 \\ \hline \$12 \end{array}$$ valves $15 other

   $$\begin{array}{r} 54 \\ 4 \\ 12 \\ +\ 15 \\ \hline \$85 \end{array}$$ total

7. 750 *lb. per day*

   $3\overline{)9000}$ = 3000 *lb. per bin*

   $4\overline{)3000}$ = 750 *lb. daily*

8. 25,000 *Btu per room* average

   $6\overline{)150{,}000}$ = 25,000 *Btu per room* average

9. 80% average grade

   $$\begin{array}{r} 75 \\ 90 \\ 87 \\ 63 \\ +\ 85 \\ \hline 400\ \% \end{array}$$

   $5\overline{)400}$ = 80 % average

10. 14,144 *sq. ft.*

   $$\begin{array}{r} 18{,}065 \\ -\ 3{,}921 \\ \hline 14{,}144\ sq.\ ft. \end{array}$$

ANSWERS TO WORKBOOK, page 5

1. 19 *studs*

   $16\overline{)288}$ = 18 *spaces* ⇒ use 18 studs

   $$\begin{array}{r} 18 \\ +\ 1 \\ \hline 19 \end{array}$$ (beginning stud) — 19 *studs* total

2. 780 *wire connectors*

   $$\begin{array}{r} 260 \\ \times\ 3 \\ \hline 780 \end{array}$$ *wire connectors* total

3. 3450 *gal.* remaining

$$\begin{array}{r} 175 \\ 192 \\ 256 \\ 314 \\ 88 \\ 153 \\ +\ 372 \\ \hline 1550 \end{array} \ \textit{gal.}\text{ delivered}$$

$$\begin{array}{r} 5000 \\ -\ 1550 \\ \hline 3450 \end{array} \ \textit{gal.}\text{ remaining}$$

4. 4 *in.*

$$\begin{array}{r} 3 \\ 2 \\ 5 \\ 1 \\ 3 \\ +\ 3 \\ \hline 17 \end{array} \ \textit{in.}\text{ given}$$

$$\begin{array}{r} 21 \\ -\ 17 \\ \hline 4 \end{array} \ \textit{in.}\text{ remaining}$$

5. (a) Job #1, (b) $192 *per wk.*

Job #1: $\begin{array}{r} 32 \\ \times\ 6 \\ \hline \$192 \end{array}$ *per wk.* Job #2: $\begin{array}{r} 24 \\ \times\ 7 \\ \hline \$168 \end{array}$ *per wk.*

6. 8 *hr.* flight time

$$16\overline{)128} = 8 \ \textit{hr.}$$

7. 544 *pills* altogether

$$\begin{array}{r} 148 \\ 66 \\ 31 \\ +\ 27 \\ \hline 272 \end{array} \ \textit{pills}\text{ before lunch}$$

$$\begin{array}{r} 272 \\ \times\ 2 \\ \hline 544 \end{array} \ \textit{pills}\text{ total}$$

8. 180 *lin. ft.* additional required

$\begin{array}{r} 23 \\ \times\ 8 \\ \hline 184 \end{array}$ *lin. ft.* $\begin{array}{r} 14 \\ \times\ 14 \\ \hline 196 \end{array}$ *lin. ft.* $\begin{array}{r} 8 \\ \times\ 11 \\ \hline 88 \end{array}$ *lin. ft.*

$$\begin{array}{r} 184 \\ 196 \\ +\ 88 \\ \hline 468 \end{array} \ \textit{lin. ft.}\text{ on-hand}$$

$$\begin{array}{r} 18 \\ \times\ 36 \\ \hline 648 \end{array} \ \textit{lin. ft.}\text{ total requirement}$$

$$\begin{array}{r} 648 \\ -\ 468 \\ \hline 180 \end{array} \ \textit{lin. ft.}\text{ additional needed}$$

9. 840 *welds* in 24 hours

$$\begin{array}{r} 35 \\ \times\ 24 \\ \hline 840 \end{array} \ \textit{welds}$$

10. 25 *60-watt light bulbs*

$$60\overline{)1500} = 25 \ \textit{bulbs}$$

11. 504 *mi.*

$$\begin{array}{r} 28 \\ \times\ 18 \\ \hline 504 \end{array} \ \textit{mi.}$$

12. $405 remains

$$\begin{array}{r} 240 \\ 580 \\ 320 \\ 290 \\ +\ 85 \\ \hline \$1515 \end{array} \ \text{expenses}$$

$$\begin{array}{r} 1920 \\ -\ 1515 \\ \hline \$405 \end{array} \ \text{remains}$$

13. 35 *in.* for dimension A

$$\begin{array}{r} 39 \\ +\ 12 \\ \hline 51 \end{array} \ \textit{in.}\text{ given}$$

$$\begin{array}{r} 86 \\ -\ 51 \\ \hline 35 \end{array} \ \textit{in.}\text{ remaining}$$

14. $158,100 estimated cost

$$\begin{array}{r} 1860 \\ \times\ 85 \\ \hline \$158{,}100 \end{array} \ \text{estimated}$$

15. 84 *cups* total in the bag

$$\begin{array}{r} 5 \\ 4 \\ +\ 3 \\ \hline 12 \end{array} \ \textit{cups}\text{ each day}$$

$$\begin{array}{r} 12 \\ \times\ 7 \\ \hline 84 \end{array} \ \textit{cups}\text{ total}$$

16. 24 *bags* of cement mix

$$3\overline{)36} = 12 \ \textit{loads}$$

$$\begin{array}{r} 12 \\ \times\ 2 \\ \hline 24 \end{array} \ \textit{bags}\text{ total}$$

17. 61 *degrees* average

$$\begin{array}{r} 87 \\ +\ 35 \\ \hline 122 \end{array} \ \textit{degrees}$$

$$2\overline{)122} = 61 \ \textit{degrees}$$

18. 24 *sheets* of plywood

4
32
24
32
+ 4
96 *ft.* total length

4 | 96 = 24 *sheets*

19. $307 total pay

3
× 9
27 *hr.* on brakes

27
× 8
$216 for brakes

1
× 13
13 *hr.* on mufflers

13
× 7
$91 for mufflers

216
+ 91
$307 total pay

20. 15 *bags* of cement

80 | 1200 = 15 *bags*

ANSWERS TO TEST MASTERS, page 80

1. 2180 W

5680
− 3500
2180 W

2. 15 boards unused

164
+ 256
420 boards total

95
+ 310
405 boards used

420
− 405
15 boards

3. 220 qt. of dirty oil

55
× 4
220 qt.

4. $9 per hour

48 | 432 = $9 per hour

5. 163 pounds after diet

178
− 15
163 lb.

6. 4 steps

8 | 32 = 4 steps

7. 416 components total

2 | 26 = 13 components per hr.

13
× 8
104 components per robot

104
× 4
416 components total

8. $6170 remaining

8500
425
60
135
+ 50
$9170 total cost

9170
− 3000
$6170 remaining

9. 581 kWh average

744
672
682
600
528
527
434
403
450
620
630
+ 682
6972 kilowatts total

12 | 6972 = 581 kWh

10. 2 min. additional

40 | 1000 = 25 gal. per min.

25 | 50 = 2 min.

Chapter 2 FRACTIONS

Objectives

After studying this chapter, the student will be able to:

- Explain basic terminology and functions of fractions.
- Reduce fractions to lowest terms.
- Find lowest common denominators.
- Perform operations of arithmetic on fractions.

ANSWERS TO TEXT, page 29

Changing Fractions From One Form to Another

1. $4\frac{5}{6}$
2. $\frac{17}{3}$
3. $\frac{9}{1}$
4. 8
5. $\frac{23}{9}$
6. $7\frac{2}{5}$
7. $8\frac{1}{6}$
8. $\frac{27}{8}$
9. $\frac{19}{4}$
10. $9\frac{1}{3}$

Equivalent Fractions

11. $\frac{\mathbf{9}}{12}$
12. $\frac{\mathbf{6}}{10}$
13. $\frac{\mathbf{72}}{12}$
14. $\frac{\mathbf{48}}{30}$
15. $3\frac{\mathbf{9}}{21}$
16. $7\frac{\mathbf{18}}{81}$
17. $\frac{\mathbf{100}}{10}$
18. $\frac{\mathbf{16}}{8}$
19. $\frac{\mathbf{9}}{48}$
20. $\frac{\mathbf{50}}{100}$
21. $\frac{1}{2}$
22. $\frac{1}{3}$
23. $\frac{1}{4}$
24. $\frac{5}{9}$
25. $1\frac{1}{2}$
26. $\frac{2}{9}$
27. 4
28. $4\frac{1}{2}$
29. $\frac{1}{10}$
30. $\frac{7}{13}$

31.	$\frac{5}{6}$	$\frac{9}{12}$	$\frac{2}{3}$	$\frac{8}{24}$	12
32.	$\frac{3}{8}$	$\frac{1}{4}$	$\frac{7}{32}$	$\frac{3}{16}$	32
33.	$\frac{3}{6}$	$\frac{35}{72}$	$\frac{7}{18}$	$\frac{11}{36}$	72
34.	$\frac{43}{54}$	$\frac{7}{9}$	$\frac{13}{18}$	$\frac{2}{3}$	54
35.	$\frac{17}{22}$	$\frac{8}{11}$	$\frac{31}{44}$	$\frac{15}{22}$	44

Addition of Fractions

36. 1
37. $1\frac{1}{3}$
38. $1\frac{3}{8}$
39. $6\frac{3}{10}$
40. $11\frac{2}{9}$
41. $4\frac{1}{8}$
42. $11\frac{13}{60}$
43. $5\frac{4}{21}$
44. $1\frac{95}{143}$
45. $5\frac{15}{32}$

Subtraction of Fractions

46. $\frac{1}{4}$
47. $\frac{1}{12}$
48. $\frac{1}{2}$

49. $1\frac{1}{2}$

50. $2\frac{3}{4}$

51. $3\frac{11}{15}$

52. $5\frac{1}{3}$

53. $2\frac{1}{6}$

54. $\frac{1}{8}$

55. $\frac{23}{30}$

Multiplication of Fractions

56. $\frac{2}{9}$

57. $\frac{3}{10}$

58. $\frac{1}{4}$

59. $\frac{1}{4}$

60. $2\frac{7}{10}$

61. 3

62. $2\frac{23}{48}$

63. $15\frac{13}{15}$

64. 48

65. $6\frac{1}{3}$

Division of Fractions

66. 2

67. $\frac{3}{8}$

68. $\frac{18}{25}$

69. $\frac{25}{36}$

70. $1\frac{1}{3}$

73. $\frac{23}{30}$

72. $4\frac{1}{4}$

73. $1\frac{31}{33}$

74. 0

75. $\frac{2}{3}$

Word Problems with Fractions

76. 84 boards are other sizes

$$147 \times \frac{3}{7} = \frac{441}{7} = 63 \text{ boards (2 x 4s)}$$

$$\begin{array}{r} 147 \\ -\ 63 \\ \hline 84 \end{array} \text{ boards are other sizes}$$

77. $8\frac{3}{4}$ hr. on the fifth day

$$6\frac{1}{3} \times 3 = \frac{19}{3} \times \frac{3}{1} = 19 \text{ hr.}$$

$$\begin{array}{r} 19 \\ +\ 12\frac{1}{4} \\ \hline 31\frac{1}{4} \end{array} \text{ hr.}$$

$$\begin{array}{rcr} 40 & = & 39\frac{4}{4} \\ -\ 3\frac{1}{4} & = & 31\frac{1}{4} \\ \hline & & 8\frac{3}{4} \end{array} \text{ hr.}$$

78. $2\frac{19}{20}$ ft.

$$14\frac{3}{4} \div 5 = \frac{59}{4} \times \frac{1}{5} = \frac{59}{20} = 2\frac{19}{20} \text{ ft.}$$

79. $6\frac{9}{10}$ yr. average age

$$\begin{array}{rcr} 8\frac{1}{4} & = & 8\frac{3}{12} \\ 6\frac{1}{2} & = & 6\frac{6}{12} \\ 2\frac{2}{3} & = & 2\frac{8}{12} \\ 9\frac{3}{4} & = & 9\frac{9}{12} \\ +\ 7\frac{1}{3} & = & 7\frac{4}{12} \\ \hline & & 32\frac{30}{12} = \frac{414}{12} \text{ yr. total} \end{array}$$

$$\frac{414}{12} \div 5 = \frac{414}{12} \times \frac{1}{5} = \frac{414}{60} = 6\frac{54}{60} = 6\frac{9}{10} \text{ yr. average}$$

80. 84¢ total cost

$$24 \times 3\frac{1}{2} = \frac{24}{1} \times \frac{7}{2} = \frac{168}{2} = 84¢$$

81. 12 in. ID

$\frac{3}{8} \times 2 = \frac{6}{8} = \frac{3}{4}$ in.

$$\begin{array}{r} 12\frac{3}{4} \\ -\ \frac{3}{4} \\ \hline 12 \text{ in.} \end{array}$$

82. $1287 \frac{1}{2}$ lb. lost

$$\begin{array}{rcccc} 54\frac{1}{4} & = & 54\frac{3}{12} & = & 53\frac{15}{12} \\ -\ 45\frac{2}{3} & = & 45\frac{8}{12} & = & 45\frac{8}{12} \\ \hline \end{array}$$

$8 \frac{7}{12}$ lb. per casting

$8\frac{7}{12} \times 150 = \frac{103}{12} \times 150 = \frac{15{,}450}{12} = 1287 \frac{1}{2}$ lb.

83. $50 \frac{1}{4}$ ft. of wire

$13 \frac{3}{4} \times 3 = \frac{55}{4} \times 3 = \frac{165}{4} = 41 \frac{1}{4}$ ft.

$\frac{18}{12} \times 3 = \frac{54}{12}$ ft. extra at one end

$\frac{54}{12} \times 2 = \frac{108}{12}$ ft. = 9 ft. extra for both ends

$$\begin{array}{r} 41\frac{1}{4} \\ +\ 9 \\ \hline 50\frac{1}{4} \text{ ft.} \end{array}$$

84. $1 \frac{7}{8}$ in. between centers

$1 \frac{7}{16} \times 2 = \frac{23}{16} \times 2 = \frac{46}{16}$ in. $= 2 \frac{7}{8}$ in.

$$\begin{array}{rcc} 10\frac{3}{8} & = & 9\frac{11}{8} \\ -\ 2\frac{7}{8} & = & 2\frac{7}{8} \\ \hline \end{array}$$

$7 \frac{4}{8}$ in. $= 7 \frac{1}{2}$ in. between A and B

$7 \frac{1}{2} \div 4 = \frac{15}{2} \times \frac{1}{4} = \frac{15}{8}$ in. $= 1\frac{7}{8}$ in.

85. $\frac{7}{32}$ in. deep cut

$\frac{7}{16} \div 2 = \frac{7}{16} \times \frac{1}{2} = \frac{7}{32}$ in.

ANSWERS TO WORKBOOK, page 11

Changing Fractions From One Form To Another

1. $\frac{11}{3}$
2. $8 \frac{1}{2}$
3. $\frac{11}{4}$
4. $6 \frac{2}{5}$
5. $\frac{10}{1}$
6. $1 \frac{4}{5}$
7. $\frac{60}{10}$
8. $15 \frac{5}{11}$
9. $\frac{481}{40}$
10. 5

Equivalent Fractions

11. $\frac{\mathbf{12}}{16}$
12. $\frac{\mathbf{57}}{95}$
13. $1 \frac{\mathbf{45}}{81}$
14. $\frac{\mathbf{10}}{32}$
15. $\frac{\mathbf{36}}{12}$
16. $4 \frac{\mathbf{3}}{18}$
17. $\frac{\mathbf{90}}{30}$
18. $\frac{\mathbf{32}}{32}$
19. $\frac{\mathbf{50}}{100}$
20. $15 \frac{\mathbf{56}}{48}$
21. $\frac{1}{2}$
22. $\frac{1}{3}$
23. $\frac{1}{7}$
24. $\frac{5}{18}$
25. $\frac{1}{2}$
26. $1 \frac{1}{7}$

27. $5\frac{1}{3}$

28. $2\frac{10}{11}$

29. $\frac{13}{25}$

30. $2\frac{1}{32}$

31. $\frac{3}{4}$ $\frac{1}{2}$ $\frac{5}{16}$ $\frac{9}{32}$ 32

32. $2\frac{3}{4}$ $2\frac{17}{24}$ $2\frac{2}{3}$ $2\frac{7}{12}$ 24

33. $\frac{7}{25}$ $\frac{1}{4}$ $\frac{23}{100}$ $\frac{11}{50}$ 100

34. $\frac{3}{5}$ $\frac{1}{2}$ $\frac{11}{25}$ $\frac{4}{10}$ 50

35. $\frac{2}{7}$ $\frac{1}{4}$ $\frac{3}{14}$ $\frac{5}{28}$ 28

36. $\frac{7}{16}$ $\frac{1}{2}$ $\frac{5}{8}$ $\frac{3}{4}$ 16

37. $\frac{5}{64}$ $\frac{3}{32}$ $\frac{1}{8}$ $\frac{3}{16}$ 64

38. $\frac{4}{10}$ $\frac{12}{25}$ $\frac{1}{2}$ $\frac{51}{100}$ 100

39. $\frac{15}{16}$ 1 $1\frac{2}{64}$ $\frac{34}{32}$ 32

40. $5\frac{3}{36}$ $5\frac{1}{4}$ $5\frac{1}{3}$ $5\frac{4}{9}$ 36

Addition of Fractions

41. 1

42. $1\frac{1}{4}$

43. $\frac{14}{15}$

44. $1\frac{7}{12}$

45. $6\frac{25}{56}$

46. $5\frac{11}{15}$

47. $1\frac{11}{12}$

48. $1\frac{1}{6}$

49. $12\frac{1}{10}$

50. $6\frac{1}{30}$

Subtraction of Fractions

51. $\frac{1}{2}$

52. $\frac{1}{2}$

53. $\frac{1}{4}$

54. $4\frac{5}{8}$

55. $1\frac{3}{4}$

56. $7\frac{1}{8}$

57. $2\frac{5}{6}$

58. $\frac{1}{2}$

59. $\frac{8}{21}$

60. 1

Multiplication of Fractions

61. $\frac{1}{6}$

62. $\frac{2}{5}$

63. $4\frac{2}{7}$

64. 15

65. $20\frac{1}{8}$

66. $5\frac{1}{2}$

67. $29\frac{1}{4}$

68. $35\frac{7}{27}$

69. $16\frac{1}{10}$

70. $\frac{1}{6}$

Division of Fractions

71. $\frac{5}{6}$

72. 1

73. 8

74. $\frac{55}{76}$

75. $\frac{7}{34}$

76. $11\frac{1}{3}$

77. 8

78. $\frac{3}{4}$

79. $\frac{1}{3}$

80. $2\frac{7}{11}$

Word Problems with Fractions

81. \$75 for savings

$$\frac{1}{4} = \frac{3}{12}$$
$$\frac{1}{3} = \frac{4}{12}$$
$$+\ \frac{1}{6} = \frac{2}{12}$$
$$\frac{9}{12} = \frac{3}{4} \text{ expenses}$$

$$300 \times \frac{3}{4} = \$225 \text{ expenses}$$

$$300 - 225 = \$75 \text{ for savings}$$

82. 800 male voters

$$2000 \times \frac{3}{5} = 1200 \text{ female}$$

$$2000 - 1200 = 800 \text{ male}$$

83. $10\frac{19}{24}$ in. each piece

$$32\frac{3}{8} \div 3 = \frac{259}{8} \times \frac{1}{3} = \frac{259}{24} \text{ in.} = 10\frac{19}{24} \text{ in.}$$

84. 8 cages

$$14 \div 1\frac{3}{4} = 14 \times \frac{4}{7} = 8 \text{ cages}$$

85. $34\frac{1}{96}$ average lb. per fish

$$22\frac{1}{2} = 22\frac{12}{24}$$
$$35\frac{2}{3} = 35\frac{16}{24}$$
$$48\frac{1}{4} = 48\frac{6}{24}$$
$$+\ 29\frac{5}{8} = 29\frac{15}{24}$$
$$134\frac{49}{24} = 136\frac{1}{24} \text{ lb. total}$$

$$136\frac{1}{24} \div 4 = \frac{3265}{24} \times \frac{1}{4} = \frac{3265}{96} = \text{average lb./fish}$$
$$= 34\frac{1}{96} \text{ average lb./fish}$$

ANSWERS TO TEST MASTERS, page 83

1. $5\frac{5}{6}$
2. $\frac{19}{5}$
3. $\frac{\mathbf{10}}{12}$
4. $\frac{\mathbf{12}}{28}$
5. $\frac{7}{18}$
6. $2\frac{1}{4}$
7. $\frac{3}{4}$ $\frac{23}{32}$ $\frac{5}{8}$ $\frac{9}{16}$ 32
8. $\frac{2}{3}$ $\frac{1}{2}$ $\frac{5}{12}$ $\frac{1}{4}$ 12
9. $\frac{34}{35}$
10. $\frac{3}{8}$
11. $\frac{1}{3}$
12. $1\frac{7}{8}$
13. $8\frac{17}{24}$
14. $1\frac{7}{12}$
15. $12\frac{7}{15}$
16. $\frac{77}{90}$
17. 5 in.

$$3\frac{1}{2} = 3\frac{4}{8}$$
$$\frac{5}{8} = \frac{5}{8}$$
$$\frac{1}{2} = \frac{4}{8}$$
$$+\ \frac{3}{8} = \frac{3}{8}$$
$$3\frac{16}{8} = 5 \text{ in.}$$

18. $15\frac{1}{8}$ ft.

$$\begin{array}{rcl} 4 & = & 4 \\ 4 & = & 4 \\ 3\frac{3}{4} & = & 3\frac{6}{8} \\ +\ 3\frac{5}{8} & = & 3\frac{5}{8} \\ \hline & & 14\frac{11}{8} = 15\frac{3}{8} \text{ ft.} \end{array}$$

$$\begin{array}{rcl} 30\frac{1}{2} & = & 30\frac{4}{8} \\ -\ 15\frac{3}{8} & = & 15\frac{3}{8} \\ \hline & & 15\frac{1}{8}\text{ft.} \end{array}$$

19. $\frac{3}{4}$ in. removed

$\frac{3}{16} \times 4 = \frac{12}{16} = \frac{3}{4}$ in.

20. 46 fixtures, 9 1/3 ft.

(a) $500 \div 10\frac{2}{3} = 500 \div \frac{32}{3} = 500 \times \frac{3}{32} =$

$\frac{1500}{32} = 46\frac{28}{32} = 46\frac{7}{8} \Rightarrow$ 46 fixtures

(b) $10\frac{2}{3} \times \frac{7}{8} = \frac{32}{3} \times \frac{7}{8} = \frac{224}{24} = 9\frac{8}{24} = 9\frac{1}{3}$ ft.

Chapter 3
DECIMAL FRACTIONS

Objectives

After studying this chapter, the student will be able to:

- Explain basic terminology and functions of decimals.
- Convert fractions to decimals.
- Convert decimals to fractions.
- Identify significant figures of numbers.
- Round off numbers to indicated degrees of accuracy or precision.
- Perform the operations of arithmetic on decimals.

ANSWERS TO TEXT, page 42

Place Values—Whole Numbers

1. 1 × 1,000
   3 × 100
   2 × 10
   4 × 1
2. 9 × 10,000
   0 × 1000
   0 × 100
   2 × 10
   7 × 1
3. 8 × 100
   1 × 10
   9 × 1
4. 2 × 1,000,000
   5 × 100,000
   6 × 10,000
   0 × 1 000
   4 × 100
   0 × 10
   9 × 1

Place Values—Decimal Numbers

5. $\frac{3}{100}$
6. $\frac{5}{1000}$
7. $\frac{79}{1,000,000}$
8. $\frac{106}{1000}$
9. $\frac{5}{10}$
10. $\frac{204}{100,000}$
11. $\frac{49}{1000}$
12. $\frac{6}{10,000}$

Rounding Numbers—Significant Figures

13. 29.1
14. 1.56
15. 0.0739
16. 87,300
17. 561
18. 3.46
19. 0.727

Rounding Numbers—Decimal Places

20. 2.500
21. 0.005
22. 29.600
23. 0.009
24. 0.036
25. 0.022

Addition and Subtraction of Decimal Numbers

26. 59.861
27. 14.6102
28. 113.06
29. 32.15
30. 0.0026
31. 0.009
32. 1060.001
33. 1.0005
34. 5591.327
35. 108.981

Multiplication and Division of Decimal Numbers

36. 97.6
37. 61.04
38. 0.0005
39. 513.525
40. 0.0053
41. 70
42. 7
43. 0.6
44. 0.25
45. 0.22
46. 1.071
47. 1872
48. 1458.333
49. 562.5
50. 127.619

Converting Decimals and Fractions

51. 0.6
52. 0.875
53. 3.8
54. 0.188
55. 1.125
56. $\frac{3}{8}$
57. $\frac{7}{10}$
58. $\frac{5}{8}$
59. $8\frac{2}{5}$
60. $4\frac{3}{100}$

Word Problems with Decimals

61. 71.1 ft.

$$\begin{array}{r} 12.3 \\ 45.68 \\ +\ 13.075 \\ \hline 71.055 \end{array} \text{ ft.}$$

62. 7.5 A

$$\begin{array}{r} 12.3 \\ -\ 4.8 \\ \hline 7.5 \end{array} \text{ A}$$

63. $89.60

$32 \times 2.8 = \$89.6$

64. 6 passes

$0.018 \div 0.003 = 6$ passes

65. 20 oz. box is the better buy

$2.30 \div 15 = \$0.153/\text{oz.}$
$2.55 \div 20 = \$0.128/\text{oz.}$

66. $276

$$\begin{array}{r} 12.3 \\ 5.4 \\ 18.45 \\ +\ 13.06 \\ \hline 49.21 \end{array} \text{ ft.}$$

$49.21 \times 5.60 = \$275.5760$

67. 257 pieces average

$$\begin{array}{r} 230 \\ 310 \\ 190 \\ 285 \\ +\ 269 \\ \hline 1284 \end{array} \text{ pieces}$$

$1284 \div 5 = 256.8$ pieces/load average

68. 2.19 hr.

$32 \div 14.6 = 2.192$ hr.

69. 0.0625 in.

$\frac{7}{16} = 0.4375$

$$\begin{array}{r} 0.4375 \\ -\ 0.3750 \\ \hline 0.0625 \end{array} \text{ in.}$$

70. 1070 gal. remaining

$$\begin{array}{r} 234.5 \\ 197.3 \\ 308.63 \\ 480.02 \\ +\ 212.35 \\ \hline 1432.80 \end{array} \text{ gal.} \qquad \begin{array}{r} 2500.00 \\ -\ 1432.80 \\ \hline 1067.20 \end{array} \text{ gal.}$$

ANSWERS TO WORKBOOK, page 23

Place Values—Whole Numbers

1. **7** × **100**
   9 × **10**
   0 × **1**
2. **3** × **1000**
   0 × **100**
   2 × **10**
   6 × **1**
3. **8** × **10,000**
   3 × **1000**
   2 × **100**
   0 × **10**
   1 × **1**
4. **4** × **100,000**
   0 × **10,000**
   1 × **1000**
   3 × **100**
   2 × **10**
   9 × **1**

Place Values—Decimal Numbers

5. $\mathbf{2 \times 10}$ $\quad$ $\mathbf{3 \times \frac{1}{10}}$

$\mathbf{5 \times 1}$ $\quad$ $\mathbf{6 \times \frac{1}{100}}$

$\mathbf{9 \times \frac{1}{1000}}$

6. $\frac{3}{10}$
7. $\frac{25}{100}$
8. $\frac{125}{1000}$
9. $\frac{50}{1000}$
10. $\frac{67}{10,000}$
11. $\frac{4}{10,000}$
12. $\frac{5}{10}$
13. $\frac{375}{1000}$
14. $\frac{81}{100}$
15. $\frac{666}{100,000}$

Rounding Numbers—Significant Figures

16. 365
17. 0.0579
18. 5.93
19. 0.116
20. 26.9
21. 0.00508
22. 0.981
23. 2.56
24. 0.0346
25. 0.599

Rounding Numbers—Decimal Places

26. 5.060
27. 0.264
28. 0.001
29. 0.025
30. 0.387
31. 0.556
32. 1.000
33. 3.075
34. 1.510
35. 2.255

Addition and Subtraction of Decimal Numbers

36. 1707.222
37. 0.7949
38. 61.59
39. 0.08231
40. 131.416
41. 0.58396
42. 353.19
43. 48.385
44. 26.565
45. 32.164

Multiplication and Division of Decimal Numbers

46. 0.4
47. 0.00177
48. 300
49. 0.375
50. 0.188
51. 0.0205
52. 400
53. 30
54. 6
55. 0.0625

Converting Decimals and Fractions

56. 0.6
57. 0.571
58. $\frac{1}{8}$
59. $\frac{7}{20}$
60. 0.438
61. 2.563
62. $3\frac{1}{4}$
63. $1\frac{1}{20}$
64. 0.4
65. 0.833

Word Problems with Decimals

66. 25.1 tons

$$\begin{array}{r} 3.75 \\ +\ 29.4 \\ \hline 33.15 \text{ tons} \end{array} \qquad \begin{array}{r} 33.15 \\ -\ 8.06 \\ \hline 25.09 \text{ tons} \end{array}$$

67. 2.91 A

$$6.8 \times \frac{3}{7} = \frac{20.4}{7} = 2.914 \text{ A}$$

68. 130 gallons

$$3200 \div 24.6 = 130.081$$

69. 1.16 cm

$$\begin{array}{r} 96.4 \\ -\ 94.08 \\ \hline 2.32 \text{ cm} \end{array} \qquad 2.32 \div 2 = 1.16 \text{ cm}$$

70. 338 mg

$$22.5 \times 15 = 337.5 \text{ mg}$$

71. 0.063 in.

$$\begin{array}{r} 0.026 \\ +\ 0.037 \\ \hline 0.063 \text{ in.} \end{array}$$

72. 170 bars total

$14 \div 0.4 = 35$ bars per side (includes 1 bar at 1 corner)

$$\begin{array}{r} 35 \\ -\ 1 \\ \hline 34 \end{array}$$ bars per side (excluding corner bar)

$34 \times 5 = 170$ bars

73. 17 2 x 4s

$249.6 \div 16 = 15.6$ spaces ⇒ use 16 studs

$$\begin{array}{r} 16 \\ +\ 1 \\ \hline 17 \end{array}$$ (beginning stud) studs total

74. (a) 7.625 in., (b) $7\frac{5}{8}$ in.

$$\begin{array}{r} 0.8 \\ +\ 0.8 \\ \hline 1.6 \text{ in.} \end{array} \qquad \begin{array}{r} 9.225 \\ -\ 1.600 \\ \hline 7.625 \text{ in.} \end{array} = 7\frac{625}{1000} \text{ in.} = 7\frac{5}{8} \text{ in.}$$

75. $29,640.00

$12.35 \times 48 = \$592.80$
$592.80 \times 50 = \$29,640.00$

ANSWERS TO TEST MASTERS, page 89

1. $\mathbf{6 \times 100}$ $\quad$ $\mathbf{3 \times \frac{1}{10}}$

   $\mathbf{2 \times 10}$ $\quad$ $\mathbf{8 \times \frac{1}{100}}$

   $\mathbf{7 \times 1}$ $\quad$ $\mathbf{5 \times \frac{1}{1000}}$

2. $\frac{7}{10}$
3. $\frac{38}{100}$
4. $\frac{255}{1000}$
5. $\frac{61}{10,000}$
6. 275
7. 95.9
8. 4.01
9. 0.125
10. 0.00387
11. 2.224
12. 0.667
13. 0.333
14. 0.438
15. 0.687
16. 5.012
17. 57.39
18. 12.13
19. 7.14
20. 0.575
21. 0.223
22. 0.2936
23. 0.085
24. 1.523
25. 38.5
26. 0.0333
27. $2\ \frac{9}{25}$
28. $\frac{3}{5}$
29. 0.563
30. 0.875
31. 0.710 in.

$$\begin{array}{r} 0.685 \\ +\ 0.025 \\ \hline 0.710 \text{ in.} \end{array}$$

32. 16.65 gal.

$$\begin{array}{r} 45 \\ \times\ 0.37 \\ \hline 16.65 \text{ gal.} \end{array}$$

33. 3.15 ft.

$12.6 \div 4 = 3.15$ ft.

Chapter 4
PERCENTAGE

Objectives

After studying this chapter, the student will be able to:

- Explain basic terminology and functions of percent.
- Express numerical values in terms of percent.
- Change percent values to decimals and fractions.
- Compute bases, rates, or portions of numbers.
- Determine maximum and minimum allowable measurements given percent tolerances.

ANSWERS TO TEXT, PAGE 48

Relationship of Percentage to Fractions and Decimals

1. 80%, 0.8
2. 0.065, $\frac{13}{200}$
3. 0.45, $\frac{9}{20}$
4. 3%, $\frac{3}{100}$
5. 62.5%, 0.625

Solving Basic Percent Problems

6. B = 280. R = 55%. P = ?
   Formula: $P = B \times R$
   Answer: P = 154
7. B = 120. R = ? P = 36.
   Formula: $R = \frac{P}{B}$
   Answer: R = 30%
8. B = ? R = 70%. P = 18.
   Formula: $B = \frac{P}{R}$
   Answer: B = 25.7

9. B = 80. R = 40%. P = ?
   Formula: P = B × R
   Answer: P = 32
10. B = 50. R = ? P = 125.
   Formula: $R = \frac{P}{B}$
   Answer: R = 250%
11. B = 90. R = ? P = 30.
   Formula: $R = \frac{P}{B}$
   Answer: R = 33.3%
12. B = ? R = 75%. P = 75.
   Formula: $B = \frac{P}{R}$
   Answer: B = 100

Word Problems with Percent

13. B = 15,000. R = 25%. P = ?
   Formula: P = B × R
   Answer: P = $3750
14. B = 60. R = ? P = 3.
   Formula: $R = \frac{P}{B}$
   Answer: R = 5%
15. B = ? R = 60%. P = 300.
   Formula: $B = \frac{P}{R}$
   B = 500 screws total
   Additional = 500 − 300
   Answer: 200 screws more
16. B = 30,000. R = ? P = 3000.
   Formula: $R = \frac{P}{B}$
   Answer: R = 10%
17. Motor 1: B = 500. R = ? P = 370.
   Formula: $R = \frac{P}{B}$
   Answer: R = 74%
   Motor 2: B = 550. R = ? P = 418
   Formula: $R = \frac{P}{B}$
   Answer: R = 76% (most efficient)
18. B = 29. R = 15%. P = ?
   Formula: P = B × R
   P = 4.35 gal. for waste
   29 + 4.35 = 33.35 gal. needed
   Answer: 33.35 ÷ 5 = 6.67 pails ⇒ 7 5-gal. pails needed
19. B = 14,000. P = ? P = 11,525.
   Formula: $R = \frac{P}{B}$
   Answer: R = 82.3% of capacity
20. 50 + 35 = 85% paper and plastic
   100 − 85 = 15% aluminum
   B = 1200. R = 15% P = ?
   Formula: P = B × R
   Answer: P = 180 ft.^{2}

Percent Tolerance

21. max. = 176 ft., min. = 144 ft.
22. max. = 493.5 ohms, min. = 446.5 ohms
23. max. = 0.255 oz., min. = 0.245 oz.
24. max. = 0.005065 in., min. = 0.004935 in.
25. max. = 0.125625 mm, min. = 0.124375 mm

ANSWERS TO WORKBOOK, page 33

Relationship of Percent to Fractions and Decimals

1. (a) 25%, (b) 0.25
2. (a) 50%, (b) $\frac{1}{2}$
3. (a) 0.333, (b) $\frac{1}{3}$
4. (a) 12.5%, (b) $\frac{1}{8}$
5. (a) 1.875, (b) $1\frac{7}{8}$
6. (a) 231.25%, (b) 2.3125
7. (a) 175%, (b) $1\frac{3}{4}$
8. (a) 0.4, (b) $\frac{2}{5}$
9. (a) 43.75%, (b) 0.4375
10. (a) 0.055, (b) $\frac{11}{200}$

Solving Basic Percent Problems

11. B = **360**. R = **65%**. P = ?
   Formula: **P = B × R**
   Substitution: **P = 360 × 0.65**
   Solution: **P = 234**
12. B = **96**. R = ? P = **32**.
   Formula: $\mathbf{R = \frac{P}{B}}$
   Substitution: $\mathbf{R = \frac{32}{96}}$
   Solution: **R = 33.3%**
13. B = **15**. R = **85%**. P = ?
   Formula: **P = B × R**
   Substitution: **P = 15 × 0.85**
   Solution: **P = 12.75**
14. B = **60**. R = **20%**. P = ?
   Formula: **P = B × R**
   Substitution: **P = 60 × 0.20**
   Solution: **P = 12**
15. B = **50**. R = ? P = **75**.
   Formula: $\mathbf{R = \frac{P}{B}}$
   Substitution: $\mathbf{R = \frac{75}{50}}$
   Solution: **R = 150%**
16. B = **85**. R = **85%**. P = ?
   Formula: **P = B × R**
   Substitution: **P = 85 × 0.85**
   Solution: **P = 72.25**

Word Problems with Percent

17. B = **65,000**. R = **25%**. P = ?
    Formula: **P = B × R**
    Substitution: **P = 65,000 × 0.25**
    Solution: **P = $16,250**
18. B = **40**. R = ? P = 40 − 8 = **32**
    Formula: $\mathbf{R} = \frac{\mathbf{P}}{\mathbf{B}}$
    Substitution: $\mathbf{R} = \frac{\mathbf{32}}{\mathbf{40}}$
    Solution: **R = 80%**
19. B = 140 + 180 = **320**. R = ? P = **140**.
    Formula: $\mathbf{R} = \frac{\mathbf{P}}{\mathbf{B}}$
    Substitution: $\mathbf{R} = \frac{\mathbf{140}}{\mathbf{320}}$
    Solution: **R = 43.75%**
20. B = **144**. R = ? P = **12**.
    Formula: $\mathbf{R} = \frac{\mathbf{P}}{\mathbf{B}}$
    Substitution: $\mathbf{R} = \frac{\mathbf{12}}{\mathbf{144}}$
    Solution: **R = 8.33%**

Percent Tolerance

21. maximum = **276 ft.**
    minimum = **184 ft.**
22. maximum = **588 Ω**
    minimum = **532 Ω**
23. maximum = **0.8755 g**
    minimum = **0.8245 g**
24. maximum = **0.0025375 in.**
    minimum = **0.0024625 in.**
25. maximum = **0.37575 mm**
    minimum = **0.37425 mm**

ANSWERS TO TEST MASTERS, page 92

1. (a) 75%, (b) 0.75
2. (a) 80%, (b) $\frac{4}{5}$
3. (a) 0.667, (b) $\frac{2}{3}$
4. (a) 37.5%, (b) 0.375
5. (a) 0.045, (b) $\frac{9}{200}$
6. 238
   B = 680. R = 35%. P = ?
   Formula: P = B × R
   Substitution: P = 680 × 0.35
   Solution: P = 238
7. 50%
   B = 36. R = ? P = 18.
   Formula: $R = \frac{P}{B}$
   Substitution: $R = \frac{18}{36}$
   Solution: R = 50%
8. 86.7
   B = ? R = 75%. P = 65.
   Formula: $B = \frac{P}{R}$
   Substitution: $B = \frac{65}{0.75}$
   Solution: B = 86.7
9. 3.2
   B = 8. R = 40%. P = ?
   Formula: P = B × R
   Substitution: P = 8 × 0.4
   Solution: P = 3.2
10. 300%
    B = 40. R = ? P = 120.
    Formula: $R = \frac{P}{B}$
    Substitution: $R = \frac{120}{40}$
    Solution: R = 300%
11. 45
    B = ? R = 200%. P = 90.
    Formula: $B = \frac{P}{R}$
    Substitution: $B = \frac{90}{2}$
    Solution: B = 45
12. $0.525
    B = 15. R = 3.5%. P = ?
    Formula: P = B × R
    Substitution: P = 15 × 0.035
    Solution: P = $0.525
13. 44.4%
    B = 180. R = ? P = 80.
    Formula: $R = \frac{P}{B}$
    Substitution: $R = \frac{80}{180}$
    Solution: R = 44.4%
14. 65%
    B = 2000. R = ? P = 1300.
    Formula: $R = \frac{P}{B}$
    Substitution: $R = \frac{1300}{2000}$
    Solution: R = 65%
15. 18.75%
    B = 8. R = ? P = 1.5.
    Formula: $R = \frac{P}{B}$
    Substitution: $R = \frac{1.5}{8}$
    Solution: R = 18.75%
16. maximum = **509.6 ft.**, minimum = **470.4 ft.**
17. maximum = **6440 Ω**, minimum = **4760 Ω**
18. maximum = **0.385 L**, minimum = **0.315 L**
19. maximum = **0.007174 in.**, minimum = **0.006426 in.**
20. maximum = **7.5225 g**, minimum = **7.4775 g**

Chapter 5
PERSONAL FINANCES

Objectives
After studying this chapter, the student will be able to:
- Explain basic terminology used with personal finance.
- Perform arithmetic with dollars and cents.
- Estimate net income.
- Determine simple or compound interest on a sum of money.
- Estimate monthly loan payments.
- Apply percentages to figure merchandise pricing.

ANSWERS TO TEXT, PAGE 62

Dollars and Cents
1. \$23.02

$$\begin{array}{r} 1.279 \\ \times \quad 18 \\ \hline \$23.022 \end{array}$$

2. \$2.10

$$\begin{array}{r} 8.95 \\ \times \quad 2 \\ \hline \$17.90 \end{array} \text{ total} \qquad \begin{array}{r} 20.00 \\ -\ 17.90 \\ \hline \$2.10 \end{array} \text{ change}$$

3. \$3.05
First 3 min. = 87¢ = \$.87

$$\begin{array}{r} 18 \\ -\ 3 \\ \hline 15 \end{array} \text{ min. @ 1/6 of 87¢}$$

$$87 \times \frac{1}{6} = 14.5¢ \text{ per min.}$$

$$\begin{array}{r} 14.5 \\ \times \quad 15 \\ \hline 217.5¢ \end{array} = \$2.18 \text{ for 15 min.}$$

$$\begin{array}{r} 2.18 \\ +\ \ .87 \\ \hline \$3.05 \end{array} \text{ total}$$

4. \$0.82 per lb.

$$3.5 \overline{)\,2.8.70} \quad = \$.82 \text{ per lb.}$$

5. \$6.62

$$\begin{array}{r} 49 \\ \times \quad 6 \\ \hline 294¢ \end{array} = \$2.94 \text{ for 8d} \qquad \begin{array}{r} 87 \\ \times \quad 12 \\ \hline 1044¢ \end{array} = \$10.44 \text{ for 16d}$$

$$\begin{array}{r} 2.94 \\ +\ 10.44 \\ \hline \$13.38 \end{array} \text{ total} \qquad \begin{array}{r} 20.00 \\ -\ 13.38 \\ \hline \$6.62 \end{array} \text{ change}$$

Gross Income, Net Income, and the Paycheck
6. \$241.54 gross pay per wk.

$$52 \overline{)\,12{,}560.000} \quad = \$241.538 \text{ gross pay per wk.}$$

7. \$19,968.00 salary per yr.

$$\begin{array}{r} 9.60 \\ \times \quad 40 \\ \hline \$384.00 \end{array} \text{ per wk.} \qquad \begin{array}{r} 384.00 \\ \times \quad 52 \\ \hline \$19{,}968.00 \end{array} \text{ salary per yr.}$$

8. \$559.04

$$\begin{array}{r} 10.75 \\ \times \quad 40 \\ \hline \$430.00 \end{array} \text{ straight-time pay} \qquad \begin{array}{r} 10.75 \\ \times \quad 1.5 \\ \hline \$16.125 \end{array} \text{ overtime per hour}$$

$$\begin{array}{r} 16.13 \\ \times \quad 8 \\ \hline \$129.04 \end{array} \text{ overtime pay} \qquad \begin{array}{r} 430.00 \\ +\ 129.04 \\ \hline \$559.04 \end{array} \text{ gross pay}$$

9. \$785.20

$$\begin{array}{r} 15.70 \\ \times \quad 0.25 \\ \hline \$3.9250 \end{array} \text{ hourly bonus} \qquad \begin{array}{r} 15.70 \\ +\ \ 3.93 \\ \hline \$19.63 \end{array} \text{ total hourly}$$

$$\begin{array}{r} 19.63 \\ \times \quad 40 \\ \hline \$785.20 \end{array} \text{ gross pay}$$

10. 23.4%

$$R = \frac{P}{B} = \frac{\text{tax}}{\text{gross}}$$

$$\text{Federal tax} = \frac{78.40}{560} = 0.14 = 14\%$$

$$\text{FICA} = \frac{40.04}{560} = 0.0715 = 7.15\%$$

$$\text{State tax} = \frac{12.54}{560} = 0.02239 = 2.24\%$$

$$\begin{array}{r} 14. \\ 7.15 \\ +\ 2.24 \\ \hline 23.39\% \end{array} = 23.4\% \text{ total}$$

11. \$493.30

$$\begin{array}{r} 560 \\ \times \ 0.15 \\ \hline \$84.00 \end{array} \text{ raise} \qquad \begin{array}{r} 560 \\ +\quad 84 \\ \hline \$644 \end{array} \text{ new gross}$$

$$\begin{array}{r} 644 \\ \times \ 0.234 \\ \hline \$150.696 \end{array} \text{ deductions} \qquad \begin{array}{r} 644.00 \\ -\ 150.70 \\ \hline \$493.30 \end{array} \text{ estimated net}$$

12. 21.8%

$$\begin{array}{r} 13.75 \\ \times \quad 40 \\ \hline \$550.00 \end{array} \text{ gross} \qquad \begin{array}{r} 550 \\ -\ 430 \\ \hline \$120 \end{array} \text{ deductions}$$

$$R = \frac{P}{B} = \frac{120}{550} = 0.218 = 21.8\%$$

13. \$8.76

$$\begin{array}{r} 7.50 \\ \times \quad 40 \\ \hline \$300.00 \text{ gross} \end{array} \qquad \begin{array}{r} 300 \\ \times\ 0.0713 \\ \hline 21.39 \text{ FICA} \end{array}$$

$$\begin{array}{r} 300.00 \\ 21.39 \\ 12.60 \\ 6.80 \\ +\quad 9.70 \\ \hline \$350.49 \text{ total weekly cost} \end{array}$$

$$\begin{array}{r} \$8.76 \text{ hourly cost} \\ 40 \overline{)350.49} \end{array}$$

14. \$646.00

$$\begin{array}{r} 5.60 \\ \times \quad 35 \\ \hline \$196.00 \text{ pay for hours} \end{array} \qquad \begin{array}{r} 15{,}000 \\ \times \quad 0.03 \\ \hline \$450.00 \text{ commission} \end{array}$$

$$\begin{array}{r} 450 \\ +\quad 196 \\ \hline \$646 \text{ gross pay} \end{array}$$

15. 25.15%

$$R = \frac{P}{B} = \frac{\text{tax}}{\text{gross}}$$

$$\text{Federal tax} = \frac{64.35}{430} = 0.1496 = 15\%$$

$$\text{FICA} = \frac{35.04}{430} = 0.0815 = 8.15\%$$

$$\text{Retirement} = \frac{8.58}{430} = 0.01995 = 2\%$$

$$\begin{array}{r} 15. \\ 8.15 \\ +\quad 2. \\ \hline 25.15\% \end{array}$$

Sales and Property Taxes

16. \$637.00

$$\begin{array}{r} 12{,}600 \\ -\quad 2800 \\ \hline \$9800 \text{ difference} \end{array} \qquad \begin{array}{r} 9800 \\ \times \quad 0.065 \\ \hline \$637.00 \text{ tax} \end{array}$$

17. 2.58¢

Vending price% = rate = 100% + 4.5% = 104.5%

$$B = \frac{P}{R} = \frac{60}{1.045} = 57.42¢$$

$$\begin{array}{r} 60.00 \\ -\ 57.42 \\ \hline 2.58¢ \text{ tax} \end{array}$$

18. 6.45%

$$\begin{array}{r} 15.50 \\ \times \quad 4 \\ \hline \$62.00 \text{ plywood} \end{array} \qquad \begin{array}{r} 66.00 \\ -\ 62.00 \\ \hline \$4.00 \text{ tax} \end{array}$$

$$R = \frac{P}{B} = \frac{4}{62} = 0.0645 = 6.45\%$$

19. \$276.00

23 mils = 2.3% = 0.023

$$\begin{array}{r} \$12{,}000 \\ \times \quad 0.023 \\ \hline \$276.00 \text{ tax} \end{array}$$

20. \$3698.00

43 mils = 4.3% = 0.043

$$\begin{array}{r} \$86{,}000 \\ \times \quad 0.043 \\ \hline \$3698.00 \text{ tax} \end{array}$$

Interest

21. \$13.86

$$4 \text{ months} = \frac{4}{12} = \frac{1}{3} = 0.33 \text{ year}$$

$$\begin{array}{r} 600 \\ \times\ 0.07 \\ \hline \$42 \text{ yearly} \end{array} \qquad \begin{array}{r} 42 \\ \times\ 0.33 \\ \hline \$13.86 \text{ in 4 mo.} \end{array}$$

22. 1.5 yr.

$$\begin{array}{r} 1500 \\ \times \quad 0.1 \\ \hline \$150 \text{ yearly} \end{array}$$

$$\begin{array}{r} 1.5 \text{ yr.} \\ 150 \overline{)225.0} \end{array}$$

23. \$304.53

$$\text{monthly} = 1 \text{ month} = \frac{1}{12} = 0.0833 \text{ year}$$

1st month:

$$\begin{array}{r} \text{yearly} \\ \text{interest} \\ 300 \\ \times \quad 0.06 \\ \hline \$18 \end{array} \qquad \begin{array}{r} \text{month} \\ \text{interest} \\ 18 \\ \times\ 0.0833 \\ \hline \$1.50 \end{array} \qquad \begin{array}{r} \text{new} \\ \text{balance} \\ 300. \\ +\quad 1.50 \\ \hline \$301.50 \end{array}$$

2nd month:

$$\begin{array}{r} \text{yearly} \\ \text{interest} \\ 301.50 \\ \times \quad 0.06 \\ \hline \$18.09 \end{array} \qquad \begin{array}{r} \text{month} \\ \text{interest} \\ 18.09 \\ \times\ 0.0833 \\ \hline \$1.51 \end{array} \qquad \begin{array}{r} \text{new} \\ \text{balance} \\ 301.50 \\ +\quad 1.51 \\ \hline \$303.01 \end{array}$$

3rd month:

$$\begin{array}{r} \text{yearly} \\ \text{interest} \\ 303.01 \\ \times \quad 0.06 \\ \hline \$18.18 \end{array} \qquad \begin{array}{r} \text{month} \\ \text{interest} \\ 18.18 \\ \times\ 0.0833 \\ \hline \$1.52 \end{array} \qquad \begin{array}{r} \text{new} \\ \text{balance} \\ 303.01 \\ +\quad 1.52 \\ \hline \$304.53 \end{array}$$

24. \$124.87

$$\text{semiannually} = 6 \text{ months} = \frac{6}{12} = 0.5 \text{ year}$$

1st 6 months:

$$\begin{array}{r} \text{yearly} \\ \text{interest} \\ 1000 \\ \times \quad 0.08 \\ \hline \$80 \end{array} \qquad \begin{array}{r} \text{6-month} \\ \text{interest} \\ 80 \\ \times \quad 0.5 \\ \hline \$40 \end{array} \qquad \begin{array}{r} \text{new} \\ \text{balance} \\ 1000 \\ +\quad 40 \\ \hline \$1040 \end{array}$$

2nd 6 months:

yearly interest	6-month interest	new balance
1040	83.20	1040.
× 0.08	× 0.5	+ 41.60
$83.20	$41.60	$1081.60

3rd 6 months:

yearly interest	6-month interest	new balance
1081.60	86.53	1081.60
× 0.08	× 0.5	+ 43.27
$86.53	$43.27	$1124.87

1124.87
− 1000.00
$124.87 interest earned

25. $2715.77

quarterly = 3 months = $\frac{3}{12}$ = 0.25 year

52 × 0.25 = 13 wk. per quarter

50
× 13
$650 deposited each quarter

1st quarter:

deposit	yearly interest	quarter interest	new balance
0.00	650	45.50	650.
+ 650.00	× 0.07	× 0.25	+ 11.38
$650.00	$45.50	$11.38	$661.38

2nd quarter:

deposit	yearly interest	quarter interest	new balance
661.38	1311.38	91.80	1311.38
+ 650.00	× 0.07	× 0.25	+ 22.95
$1311.38	$91.80	$22.95	$1334.33

3rd quarter:

deposit	yearly interest	quarter interest	new balance
1334.33	1984.33	138.90	1984.33
+ 650.00	× 0.07	× 0.25	+ 34.73
$1984.33	$138.90	$34.73	$2019.06

4th quarter:

deposit	yearly interest	quarter interest	new balance
2019.06	2669.06	186.83	2669.06
+ 650.00	× 0.07	× 0.25	+ 46.71
$2669.06	$186.83	$46.71	$2715.77

Estimating Monthly Loan Payments

26. $340.00

yearly interest	total interest	total cost
12,000	1080	12,000
× 0.09	× 4	+ 4320
$1080	$4320	$16,320

4 years = 48 months

$340 per mo.
48 ⟌16,320

27. $161.00

yearly interest	total interest	total cost
4000	600	4000
× 0.15	× 3	+ 1800
$600	$1800	$5800

36 months = 3 years

$161 per mo.
36 ⟌5800

28. $109.00

yearly interest	total interest	total cost
600	108	600
× 0.18	× 0.5	+ 54
$108	$54	$654

1/2 year = 0.5 year = 6 months

$109 per mo.
6 ⟌654

29. $104.00

yearly interest	total interest	total cost
2500	250	2500
× 0.10	× 2.5	+ 625
$250	$625	$3125

2.5 years = 30 months

$104 per mo.
30 ⟌3125

30. $183.00

yearly interest	total interest	total cost
8000	600	8000
× 0.075	× 5	+ 3000
$600	$3000	$11,000

5 years = 60 months

$183 per mo.
60 ⟌11,000

Merchandise Pricing Using Percentage

31. 50%

Base = $10. Portion = $15. Find rate (retail %).

$R = \frac{P}{B} = \frac{15}{10} = 1.5 = 150\%$

markup % = retail % − 100%
= 150% − 100% = 50%

32. $31.50
retail % = markup % + 100%
= 75% + 100% = 175%
Base = $18. Rate = 175%. Find portion (retail $).
P = B × R = 18 × 1.75 = $31.50

33. \$69.44
retail % = markup % + 100%
= 80% + 100% = 180%
Portion = \$125. Rate = 180%. Find base (wholesale \$).
$B = \frac{P}{R} = \frac{125}{1.80} = \$69.44$

34. \$56.58
retail % = markup % = 100%
= 130% + 100% = 230%
Base = \$24.60. Rate = 230%. Find portion (retail \$).
P = B × R = 24.60 × 2.3 = \$56.58

35. 257%
Portion = \$12.50. Base = \$3.50.
Find rate (retail %).
$R = \frac{P}{B} = \frac{12.50}{3.50} = 3.57 = 357\%$
markup % = retail % − 100%
= 357% − 100% = 257%

36. \$38.88
sale % = 100% − discount %
= 100% − 40% = 60%
Base = \$64.80. Rate = 60%. Find portion (sale \$).
P = B × R = 64.80 × 0.60 = \$38.88

37. \$89.00
sale % = 100% − discount %
= 100% − 50% = 50%
Portion = \$44.50. Rate = 50%. Find base (retail \$).
$B = \frac{P}{R} = \frac{44.50}{0.50} = \$89.00$

38. \$107.95
sale % = 100% − discount %
= 100% − 33.3% = 66.7%
Portion = \$72. Rate = 66.7%. Find base (retail \$).
$B = \frac{P}{R} = \frac{72}{0.667} = \$107.95$

39. 47.4%
Base = \$1.90. Portion = \$1.00. Find rate (sale %).
$R = \frac{P}{B} = \frac{1.00}{1.90} = 0.526 = 52.6\%$
discount % = 100% − sale %
= 100% − 52.6% = 47.4%

40. 40%
Base = \$50. Portion = 30. Find rate (sale %).
$R = \frac{P}{B} = \frac{30}{50} = 0.6 = 60\%$
discount % = 100% − sale %
= 100% − 60% = 40%

ANSWERS TO WORKBOOK, page 39

Dollars and Cents

1. \$104.39

$20.5\overline{)1800.0\,0}$ = 87.8 gal.

87.8 × 1.189 = \$104.3942 total cost

2. \$15.87

27.5 × 15 = 412.5¢ = \$4.13

20.00 − 4.13 = \$15.87 change

3. \$6.40

First 5 min. = \$1.60 = 160¢

35 − 5 = 30 min. @ 1/10 of 160¢

$160 \times \frac{1}{10} = 16¢$ per min.

16 × 30 = 480¢ = \$4.80 for 30 min.

4.80 + 1.60 = \$6.40 total

4. \$1.63 per lb.

$5.5\overline{)8.9\,500}$ = \$1.627 per lb.

5. \$1.25

18 − 2 = 16 wheels

$4\overline{)16}$ = 4 axles

4 + 1 = 5 axles

0.75 × 5 = \$3.75 toll

5.00 − 3.75 = \$1.25 change

Gross Income, Net Income, and the Paycheck

6. \$923.08 gross pay per wk.

$52\overline{)48{,}000.000}$ = \$923.077 gross pay per wk.

7. \$26,728.00 income per yr.

12.85 × 40 = \$514.00 per wk.

514.00 × 52 = \$26,728.00 income per yr.

8. \$488.80

9.40 × 40 = \$376.00 straight-time pay

9.40 × 1.5 = \$14.10 per hour overtime

14.10 × 8 = \$112.80 overtime pay

376.00 + 112.80 = \$488.80 gross pay

9. \$560.05

10.25 × 0.15 = \$1.538 hourly bonus

10.25 + 1.54 = \$11.79 total hourly

11.79 × 40 = \$471.60 straight time pa

11.79 × 1.5 = \$17.69 per hour overtime

17.69 × 5 = \$88.45 overtime pay

471.60 + 88.45 = \$560.05 gross pay

10. 33%

$R = \frac{P}{B} = \frac{\text{tax}}{\text{gross}}$

Federal tax $= \frac{57.60}{480} = 0.12 = 12\%$

FICA $= \frac{67.20}{480} = 0.14 = 14\%$

State tax $= \frac{33.60}{480} = 0.07 = 7\%$

$$\begin{array}{r} 12 \\ 14 \\ +\ 7 \\ \hline 33\% \end{array}$$ total

11. $347.33

$$\begin{array}{r} 480 \\ \times\ 0.08 \\ \hline \$38.40 \end{array}$$ raise

$$\begin{array}{r} 480.00 \\ +\ 38.40 \\ \hline \$518.40 \end{array}$$ new gross

$$\begin{array}{r} 518.40 \\ \times\ .33 \\ \hline \$171.072 \end{array}$$ deductions

$$\begin{array}{r} 518.40 \\ -\ 171.07 \\ \hline \$347.33 \end{array}$$ estimated net

12. 34%

$$\begin{array}{r} 18.50 \\ \times\ 40 \\ \hline \$740.00 \end{array}$$ gross

$$\begin{array}{r} 740.00 \\ -\ 488.40 \\ \hline \$251.60 \end{array}$$ deductions

$R = \frac{P}{B} = \frac{251.60}{740.00} = 0.34 = 34\%$

13. $9.67 per hr.

$$\begin{array}{r} 7.75 \\ \times\ 40 \\ \hline \$310.00 \end{array}$$ gross

$$\begin{array}{r} 310 \\ \times\ 0.122 \\ \hline \$37.82 \end{array}$$ FICA

$$\begin{array}{r} 310.00 \\ 37.82 \\ 18.50 \\ 4.90 \\ +\ 15.70 \\ \hline \$386.92 \end{array}$$ total weekly cost

$40\overline{)386.920} = \$9.673$ per hr.

14. $2604.00

$$\begin{array}{r} 6.80 \\ \times\ 30 \\ \hline \$204 \end{array}$$ pay for hours

$$\begin{array}{r} 120{,}000 \\ \times\ 0.02 \\ \hline \$2400 \end{array}$$ commission

$$\begin{array}{r} 2400 \\ +\ 204 \\ \hline \$2604 \end{array}$$ gross pay

15. 21.66%

$R = \frac{P}{B} = \frac{\text{tax}}{\text{gross}}$

Federal tax $= \frac{41.00}{368.00} = 0.111 = 11.1\%$

FICA $= \frac{28.50}{368.00} = 0.0774 = 7.74\%$

Retirement $= \frac{10.40}{368.00} = 0.0282 = 2.82\%$

$$\begin{array}{r} 11.1 \\ 7.74 \\ +\ 2.82 \\ \hline 21.66\% \end{array}$$

Sales and Property Taxes

16. $1095.88

$$\begin{array}{r} 35{,}000 \\ -\ 15{,}075 \\ \hline \$19{,}925 \end{array}$$ difference

$$\begin{array}{r} 19{,}925 \\ \times\ 0.055 \\ \hline \$1095.875 \end{array}$$ tax

17. 5.23¢

Vending price % = rate = 100% + 7.5% = 107.5%

$B = \frac{P}{R} = \frac{75}{1.075} = 69.77¢$

$$\begin{array}{r} 75.00 \\ 69.77 \\ \hline 5.23¢ \end{array}$$ tax

18. 3.5%

$$\begin{array}{r} 25.95 \\ \times\ 4 \\ \hline \$103.80 \end{array}$$ tapes

$$\begin{array}{r} 107.43 \\ -\ 103.80 \\ \hline \$3.63 \end{array}$$ tax

$R = \frac{P}{B} = \frac{3.63}{103.80} = 0.0349 = 3.5\%$

19. $255.00

30 mils = 3% = 0.03

$$\begin{array}{r} 8500 \\ \times\ 0.03 \\ \hline \$255.00 \end{array}$$ tax

20. 13.3 mils

$R = \frac{P}{B} = \frac{1200}{90{,}000} = 0.0133 = 1.33\% = 13.3$ mils

Interest

21. $4.95

4 months $= \frac{4}{12} = \frac{1}{3} = 0.33$ year

$$\begin{array}{r} 300 \\ \times\ 0.05 \\ \hline \$15 \end{array}$$ yearly

$$\begin{array}{r} 15 \\ \times\ 0.33 \\ \hline \$4.95 \end{array}$$ earned in 4 mo.

22. 2 yr.

$$\begin{array}{r} 2300 \\ \times\ 0.075 \\ \hline \$172.50 \end{array}$$ yearly

$172.50.\overline{)345.00.0} = 2.0$ yr.

23. $849.09

quarterly = 3 months = $\frac{3}{12}$ = 0.25 year

1st quarter:

yearly interest	quarter interest	new balance
800	48	800
× 0.06	× 0.25	+ 12
$48	$12	$812

2nd quarter:

yearly interest	quarter interest	new balance
812	48.72	812.00
× 0.06	× 0.25	+ 12.18
$48.72	$12.18	$824.18

3rd quarter:

yearly interest	quarter interest	new balance
824.18	49.45	824.18
× 0.06	× 0.25	+ 12.36
$49.45	$12.36	$836.54

4th quarter:

yearly interest	quarter interest	new balance
836.54	50.19	836.54
× 0.06	× 0.25	+ 12.55
$50.19	$12.55	$849.09

24. $97.92

semiannually = 6 months = $\frac{6}{12}$ = 0.5 year

1st 6 months:

yearly interest	6-month interest	new balance
1200	96	1200
× 0.08	× 0.5	+ 48
$96	$48	$1248

2nd 6 months:

yearly interest	6-month interest	new balance
1248	99.84	1248.
× 0.08	× 0.5	+ 49.92
$99.84	$49.92	$1297.92

1297.92
− 1200.00
$97.92 interest earned

25. $3238.77

quarterly = 3 months = $\frac{3}{12}$ = 0.25 year

52 × 0.25 = 13 wk. per quarter

60
× 13
$780 deposited each quarter

1st quarter:

quarter deposit	yearly interest	quarter interest	new balance
0.00	780	46.80	780.
+ 780.00	× 0.06	× 0.25	+ 11.70
$780.00	$46.80	$11.70	$791.70

2nd quarter:

quarter deposit	yearly interest	quarter interest	new balance
791.70	1571.70	94.30	1571.70
+ 780	× 0.06	× 0.25	+ 23.58
$1571.70	$94.30	$23.58	$1595.28

3rd quarter:

quarter deposit	yearly interest	quarter interest	new balance
1595.28	2375.28	142.52	2375.28
+ 780.00	× 0.06	× 0.25	+ 35.63
$2375.28	$142.52	$35.63	$2410.91

4th quarter:

quarter deposit	yearly interest	quarter interest	new balance
2410.91	3190.91	195.45	3190.91
+ 780.00	× 0.06	× 0.25	+ 47.86
$3190.91	$191.45	$47.86	$3238.77

Estimating Monthly Loan Payments

26. $267

yearly interest	total interest	total cost
15,600	1638	15,600
× 0.105	× 10	+ 16,380
$1638	$16,380	$31,980

10 years = 120 months

$266.50 per mo.
120 ⟌ 31,980.00

27. $150

yearly interest	total interest	total cost
6200	558	6200
× 0.09	× 5	+ 2790
$558	$2790	$8990

5 years = 60 months

$149.83 per mo.
60 ⟌ 8990.00

28. $91

yearly interest	total interest	total cost
800	23.20	800.
× 0.029	× 0.75	+ 17.40
$23.20	$17.40	$817.40

3/4 year = 0.75 year = 9 months

$90.82 per mo.
9 ⟌ 817.40

29. $117
24 months = 2 years

yearly interest	total interest	total cost
2500	157.50	2500
× 0.063	× 2	+ 315
$157.50	$315.00	$2815

$117.29 per mo.
24 ⟌2815.00

30. $210.67
8 months = 0.667 year

yearly interest	total interest	total cost
1600	128	1600.
× 0.08	× 0.667	+ 85.38
$128	$85.38	$1685.38

$210.67 per mo.
8 ⟌1685.38

Merchandise Pricing Using Percentage

31. 120%
Portion = $275. Base = $125. Find rate (retail %).
$R = \frac{P}{B} = \frac{275}{125} = 2.2 = 220\%$
markup % = retail % − 100%
= 220% − 100% = 120%

32. $0.72
retail % = markup % + 100%
= 85% + 100% = 185%
Base = $0.39. Rate = 185%. Find portion (retail $).
P = B × R = 0.39 × 1.85 = $0.72

33. $1.96
retail % = markup % + 100%
= 130% + 100% = 230%
Portion = $4.50. Rate = 230%. Find base (wholesale $).
$B = \frac{P}{R} = \frac{4.50}{2.30} = \$1.96$

34. 460%
Portion = $1.30. Base = 23.1¢. Find rate (retail %).
$R = \frac{P}{B} = \frac{1.30}{0.231} = 5.6 = 560\%$
markup % = retail % − 100%
= 560% − 100% = 460%

35. $7.88
retail % = markup % + 100%
= 50% + 100% = 150%
Base = $5.25. Rate = 150%. Find portion (retail $).
P = B × R = 5.25 × 1.5 = $7.875

36. $1.13
sale % = 100% − discount %
= 100% − 20% = 80%
Portion = $0.90. Rate = 80%. Find base (retail $).
$B = \frac{P}{R} = \frac{0.90}{0.80} = \$1.125$

37. 37.9%
Base = $16.50. Portion = $10.25. Find rate (sale %).
$R = \frac{P}{B} = \frac{10.25}{16.50} = 0.621 = 62.1\%$
discount % = 100% − sale %
= 100% − 62.1% = 37.9%

38. 28%
Base = $0.25. Portion = $0.18. Find rate (sale %).
$R = \frac{P}{B} = \frac{0.18}{0.25} = 0.72 = 72\%$
discount % = 100% − sale %
= 100% − 72% = 28%

39. $1.30
sale % = 100% − discount %
= 100% − 50% = 50%
Base = $2.60. Rate = 50%. Find portion (sale $).
P = B × R = 2.60 × 0.5 = $1.30

40. $567.16
sale % = 100% − discount %
= 100% − 33% = 67%
Portion = $380. Rate = 67%. Find base (retail $).
$B = \frac{P}{R} = \frac{380}{0.67} = \$567.16$

ANSWERS TO TEST MASTERS, page 97

1. $5.20

1.48	20.00
× 10	− 14.80
$14.80	$5.20

2. $426.40

8.20	8.20
× 40	× 1.5
$328 straight-time pay	$12.30 per hour overtime
12.30	328.00
× 8	+ 98.40
$98.40 overtime pay	$426.40 gross pay

3. $446.60

580	580
× 0.10	+ 58
$58 raise	$638 new gross
638	638.00
× 0.30	− 191.40
$191.40 deductions	$446.60 new net

4. $28.88
385
× 0.075
$28.875

5. 437.50
12,500
× 0.035
$437.50

6. $54.00
9 months $= \frac{9}{12} = \frac{3}{4} = 0.75$ year

800	72
× .09	× 0.75
$72 yearly	$54 earned in 9 mo.

7. \$642.74

 1st half year:

yearly interest	half-year interest	new balance
600	42	600
× 0.07	× 0.5	+ 21
\$42	\$21	\$621

 2nd half year:

yearly intrest	half-year interest	new balance
621	43.47	621.00
× 0.07	× 0.5	+ 21.74
\$43.47	\$21.74	\$642.74

8. \$124

 48 months = 4 years

yearly interest	total interest	total cost
4500	360	4500
× 0.08	× 4	+ 1440
\$360	\$1440	\$5940

 $48\overline{)5940.00}$ = \$123.75 per mo.

9. 60%

 Portion = \$40. Base = \$25. Find rate (retail %).

 $R = \frac{P}{B} = \frac{40}{25} = 1.6 = 160\%$

 markup % = retail % − 100%

 = 160% − 100% = 60%

10. \$24.50

 sale % = 100% − discount %

 = 100% − 30% = 70%

 Base = \$35. Rate = 70%. Find portion (sale \$).

 P = B × R = 35 × 0.7 = \$24.50

Chapter 6
GRAPHS

Objectives

After studying this chapter, the student will be able to:

- Explain basic terminology of graphs.
- Read line, bar, and circle graphs.
- Prepare line, bar, and circle graphs from tabular data.

ANSWERS TO TEXT, page 71

Line Graphs

1. 625 $\overline{\text{M}}$\$ ($\overline{\text{M}}$\$ = million \$)

$$\begin{array}{r} 100 \\ 125 \\ 75 \\ 50 \\ 75 \\ +\ 200 \\ \hline 625 \end{array}$$

 625 $\overline{\text{M}}$\$ total sales

2. 104 $\overline{\text{M}}$\$

 625 ÷ 6 = 104.167 $\overline{\text{M}}$\$ average

3. 48%

 Base = 104 $\overline{\text{M}}$\$. Portion = 50 $\overline{\text{M}}$\$. Find rate.

 $R = \frac{P}{B} = \frac{50}{104} = 0.48 = 48\%$ of average.

4. 52%

$$\begin{array}{r} 104 \\ -\ 50 \\ \hline 54 \end{array}$$

 54 $\overline{\text{M}}$\$ drop from average

 Base = 104 $\overline{\text{M}}$\$. Portion = 54 $\overline{\text{M}}$\$. Find rate.

 $R = \frac{P}{B} = \frac{54}{104} = 0.52 = 52\%$ drop from average

 Also:
$$\begin{array}{r} 100 \\ -\ 48 \\ \hline 52\% \end{array}$$
 52% drop from average

5. 370 $\overline{\text{M}}$\$

 retail % = markup % + 100%

 = 85% + 100% = 185%

 Base = 200 $\overline{\text{M}}$\$. Rate = 185%. Find portion (retail

 P = B × R = 200 × 1.85 = 370 $\overline{\text{M}}$\$

6. 37.5 $\overline{\text{M}}$\$

 Base = 125 $\overline{\text{M}}$\$. Rate = 30%. Find portion.

 P = B × R = 125 × 0.30 = 37.5 $\overline{\text{M}}$\$

7.

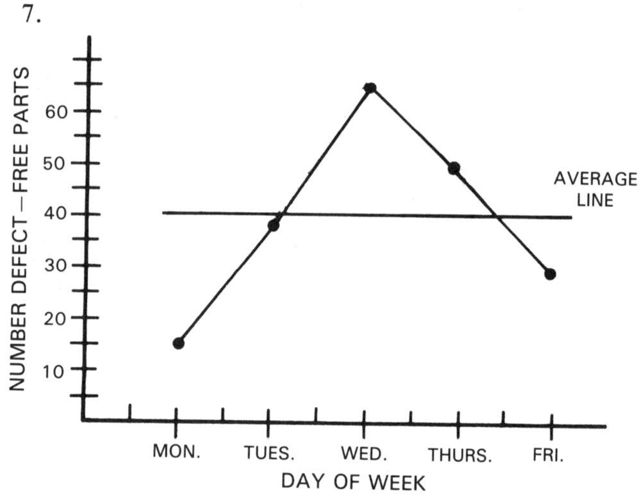

8. 198 parts

$$\begin{array}{r} 15 \\ 38 \\ 65 \\ 50 \\ +\ 30 \\ \hline 198 \end{array}$$

 198 defect-free parts

9. 39.6 parts average. Refer to Question 7 above.

 198 ÷ 5 = 39.6 parts average

10. 2

11. 37.9%

 Base = 39.6 parts. Portion = 15 parts. Find rate.

 $R = \frac{P}{B} = \frac{15}{39.6} = 0.379 = 37.9\%$

12. \$268.75 gross pay

65.0	50.0	25.4
−39.6	−39.6	+10.4
25.4 pieces	10.4 pieces	35.8 pieces

35.8 ⇒ 35 pieces above average
35 × 0.25 = $8.75 bonus
6.50 × 40 = $260.00 total hourly pay

```
   8.75
+260.00
-------
$268.75 gross pay
```

Bar Graphs

13. 2
14. 163 M$ (M$ = thousand $)

```
   20
   40
   25
    8
   35
+  35
-----
  163 M$ income
```

15. 160 M$

```
   25
   15
   35
   10
   50
+  25
-----
  160 M$ expenses
```

16.
```
  163
- 160
-----
    3 M$ profit
```

17. 62.5%

```
  40
- 15
----
  25 M$ profit
```

Base = 40 M$. Portion = 25 M$. Find rate.

$R = \frac{P}{B} = \frac{25}{40} = 0.625 = 62.5\%$ profit

18.

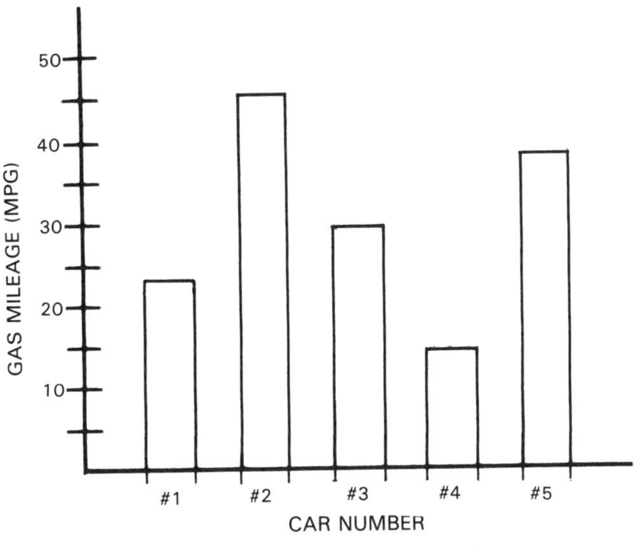

19. Car number is independent variable.
20. Gas mileage is dependent variable.
21. Car #2.
22. $25
    600 ÷ 30 = 20 gal.
    20 × 1.25 = $25 total cost
23. Car #4

Circle Graphs

24. $300
    Base = $1200. Rate = 25%. Find portion.
    P = B × R = 1200 × 0.25 = $300 rent
25. $120
    Base = $1200. Rate = 10%. Find portion.
    P = B × R = 1200 × 0.10 = $120 car loan
26. 5%
    Base = $1200. Portion = $60. Find rate.
    $R = \frac{P}{B} = \frac{60}{1200} = 0.05 = 5\%$ miscellaneous
27. 10%

```
   25           100
   10         -  90
   15         -----
   30           10% contingency
    5
+   5
-----
  90%
```

28. $120
    Base = $1200. Rate = 10%. Find portion.
    P = B × R = 1200 × 0.10 = $120 contingency
29.

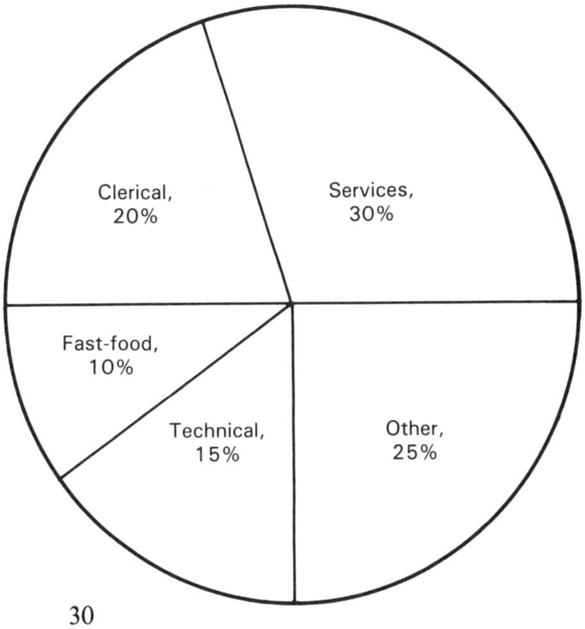


```
   30
   20
   10           100
+  15         -  75
-----         -----
  75%           25% other
```

Services: P = B × R = 360° × 0.30 = 108°
Clerical: P = B × R = 360° × 0.20 = 72°
Fast-food: P = B × R = 360° × 0.10 = 36°
Technical: P = B × R = 360° × 0.15 = 54°

30. 120 technical jobs
    Base = 800 jobs. Rate = 15%. Find portion.
    P = B × R = 800 × 0.15 = 120 jobs

31. 80 clerical jobs unfilled
    Base = 800 jobs. Rate = 20%. Find portion.
    P = B × R = 800 × 0.20 = 160 jobs

$$\begin{array}{r} 160 \\ -\ 80 \\ \hline 80 \text{ jobs} \end{array}$$

32. $22,400 per week
    40 × 7 = $280 per person per week
    280 × 80 = $22,400 per week
33. 25%
34. 200 *other* jobs
    Base = 800. Rate = 25%. Find portion.
    P = B × R = 800 × 0.25 = 200 jobs

ANSWERS TO WORKBOOK, page 47

Line Graphs

1. 50 ft.
2. 120 ft.
3. 60 mph
4. 150 ft.
5. 44 mph
6.

A STUDENT'S GRADES FOR LAB PROJECTS

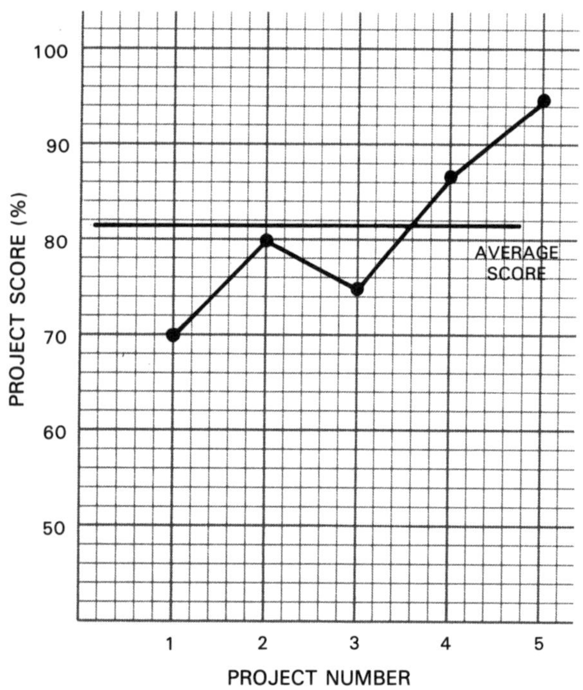

7. 81.4% (Refer to Question 6.)

$$\begin{array}{r} 70 \\ 80 \\ 75 \\ 87 \\ +\ 95 \\ \hline 407\% \end{array}$$

407 ÷ 5 = 81.4%/project

8. upward, or positive
9. 2 projects
10. 2 projects
11. 0.6 percentage points

$$\begin{array}{r} 82.0 \\ -\ 81.4 \\ \hline 0.6 \text{ points} \end{array}$$

12. 86.8%

$$\begin{array}{r} 88 \\ 88 \\ 86 \\ 86 \\ +\ 86 \\ \hline 434\ \% \end{array}$$

434 ÷ 5 = 86.8%/test

RELATIONSHIP OF ONE STUDENT'S TEST SCORES COMPARED TO THE CLASS AVERAGE

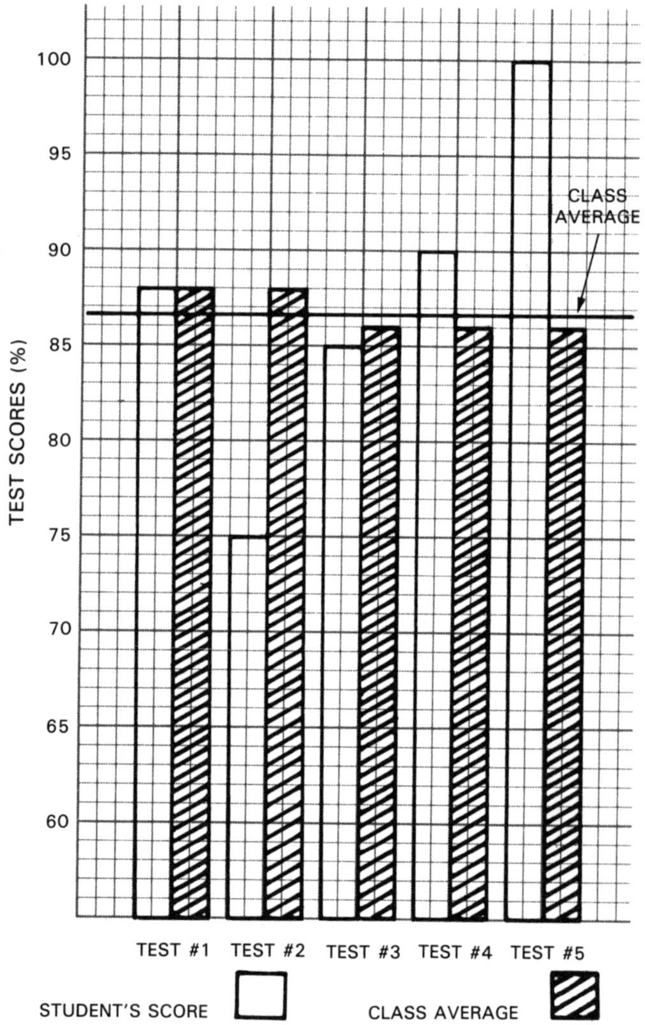

13. 87.6%

$$\begin{array}{r} 88 \\ 75 \\ 85 \\ 90 \\ +\ 100 \\ \hline 438\ \% \end{array}$$

438 ÷ 5 = 87.6%/test

14. 2 tests
15. Test #2
16. upward, or positive

17.

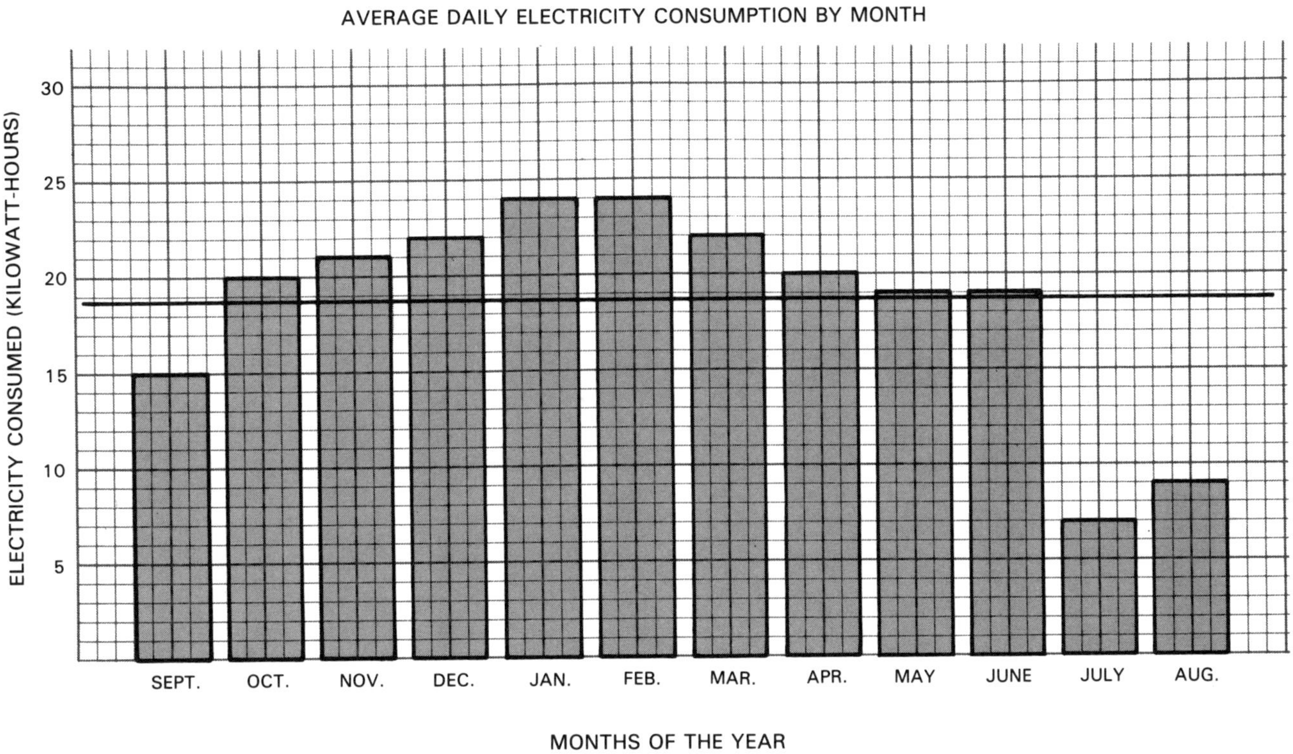

18. 18.9 kWh (Refer to Question 17.)

$$\begin{array}{r} 15 \\ 20 \\ 21 \\ 22 \\ 24 \\ 24 \\ 22 \\ 20 \\ 19 \\ 19 \\ 7 \\ +\ 14 \\ \hline 227 \end{array} \text{ kWh}$$

$227 \div 12 = 18.9$ kWh/mo.

19. (a) Jan. (b) Feb.
20. July
21. \$54.62
    $21 \times 0.074 = \$1.554$/day
    $1.554 \times 30 = \$46.62$

$$\begin{array}{r} 46.62 \\ +\ \ 8.00 \\ \hline \$54.62 \end{array} \text{ November}$$

22. \$24.06
    $7 \times 0.074 = \$0.518$/day
    $0.518 \times 31 = \$16.058$

$$\begin{array}{r} 16.06 \\ 8.00 \\ \hline \$24.06 \end{array} \text{ July}$$

23. 10%

$$\begin{array}{r} 45 \\ 5 \\ 25 \\ +\ 15 \\ \hline 90\% \end{array} \qquad \begin{array}{r} 100 \\ -\ \ 90 \\ \hline 10\% \end{array} \text{ miscellaneous}$$

24. \$1.62 million
    Base = 3.6 $\overline{M\$}$. Rate = 45%. Find portion.
    $P = B \times R = 3.6\ \overline{M\$} \times 0.45 = 1.62\ \overline{M\$}$
25. \$27,000
    Base = 3.6 $\overline{M\$}$. Rate = 15%. Find portion.
    $P = B \times R = 3.6\ \overline{M\$} \times 0.15 = 0.54\ \overline{M\$}$
    $0.54 \times \frac{1}{2} = 0.27\ \overline{M\$}$
    $0.27 \div 10 = 0.027\ \overline{M\$} = \$27{,}000$/person
26. \$120 million
    Portion = 3.6 $\overline{M\$}$. Rate = 3%. Find base.
    $B = \frac{P}{R} = \frac{3.6\ \overline{M\$}}{0.03} = 120\ \overline{M\$}$
27. \$90,000
    Base = 3.6 $\overline{M\$}$. Rate = 25%. Find portion.
    $P = B \times R = 3.6\ \overline{M\$} \times 0.25 = 0.9\ \overline{M\$}$
    Base = 0.9 $\overline{M\$}$. Rate = 10%. Find portion.
    $P = B \times R = 0.9\ \overline{M\$} \times 0.1 = 0.09\ \overline{M\$}$

28.

MONTHLY DISTRIBUTION OF EARNINGS

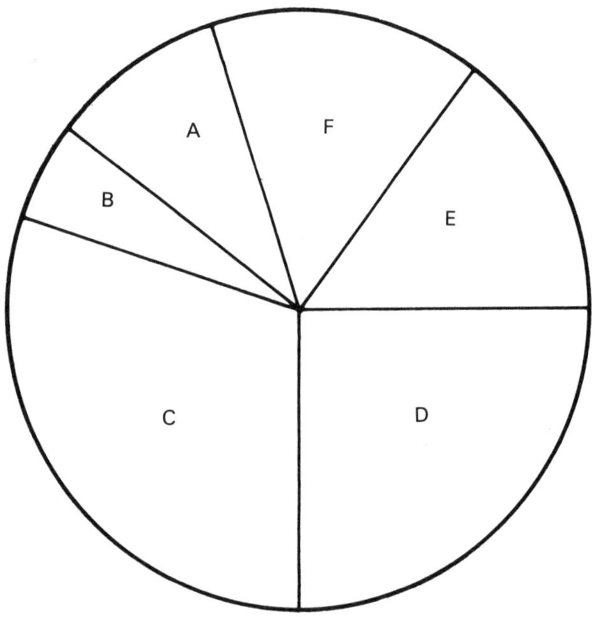

Find percentages.

A: Portion = \$5000. Base = \$50,000. Find rate.

$$R = \frac{P}{B} = \frac{\$5000}{\$50{,}000} = 0.1 = 10\%$$

E: Portion = \$7500. Base = \$50,000. Find rate.

$$R = \frac{P}{B} = \frac{\$7500}{\$50{,}000} = 0.15 = 15\%$$

F:

$$\begin{array}{r} 10 \\ 5 \\ 30 \\ 25 \\ +\ 15 \\ \hline 85\% \end{array} \qquad \begin{array}{r} 100 \\ -\ 85 \\ \hline 15\% \end{array}$$

Find angles.

A: Base = 360°. Rate = 10%. Find portion.
P = B × R = 360° × 0.1 = 36°

B: Base = 360°. Rate = 5%. Find portion.
P = B × R = 360° × 0.05 = 18°

C: Base = 360°. Rate = 30%. Find portion.
P = B × R = 360° × 0.3 = 108°

D: Base = 360°. Rate = 25%. Find portion.
P = B × R = 360° × 0.25 = 90°

E: Base = 360°. Rate = 15%. Find portion.
P = B × R = 360° × 0.15 = 54°

29. 10%

30. 15%

31. \$7500

$$\begin{array}{r} 10 \\ 5 \\ 30 \\ 25 \\ +\ 15 \\ \hline 85\% \end{array} \qquad \begin{array}{r} 100 \\ -\ 85 \\ \hline 15\% \end{array}$$

Base = \$50,000. Rate = 15%. Find portion.
P = B × R = \$50,000 × 0.15 = \$7500

32. \$75,000

Retail % = markup % + 100%
= 50% + 100% = 150%
Base = \$50,000. Rate = 150%. Find portion.
P = B × R = \$50,000 × 1.5 = \$75,000

33. \$225

$$\begin{array}{r} 10 \\ 5 \\ +\ 30 \\ \hline 45\% \end{array}$$

Base = \$50,000. Rate = 45%. Find portion.
P = B × R = \$50,000 × 0.45 = \$22,500
Base = \$22,500. Rate = 1%. Find portion.
P = B × R = \$22,500 × 0.01 = \$225

ANSWERS TO TEST MASTERS, page 100

Graphs

1. 115 cars projected

$$\begin{array}{r} 10 \\ 10 \\ 15 \\ 20 \\ 30 \\ +\ 30 \\ \hline 115 \text{ cars} \end{array}$$

2. 117 cars actual

$$\begin{array}{r} 8 \\ 15 \\ 18 \\ 25 \\ 23 \\ +\ 28 \\ \hline 117 \text{ cars} \end{array}$$

3. 7 cars down

$$\begin{array}{r} 30 \\ -\ 23 \\ \hline 7 \text{ cars} \end{array}$$

4. 5 cars over

$$\begin{array}{r} 15 \\ -\ 10 \\ \hline 5 \text{ cars} \end{array}$$

5.

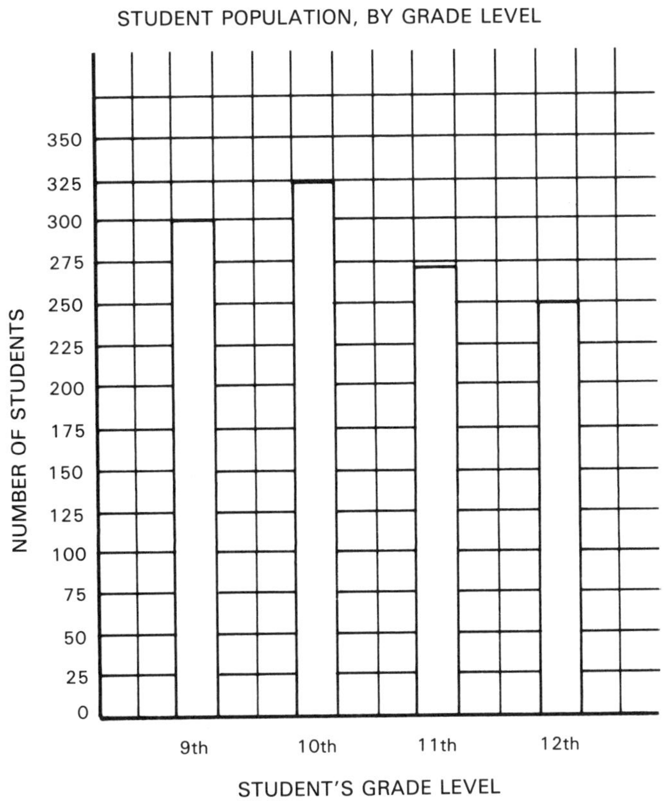

6. upward, or positive
7. Jan., Feb.
8. 2050 houses

$$\begin{array}{r} 200 \\ 200 \\ 300 \\ 450 \\ 500 \\ +\ 400 \\ \hline 2050 \text{ houses} \end{array}$$

9. \$307,500,000
   150,000 × 2050 = \$307,500,000
10.

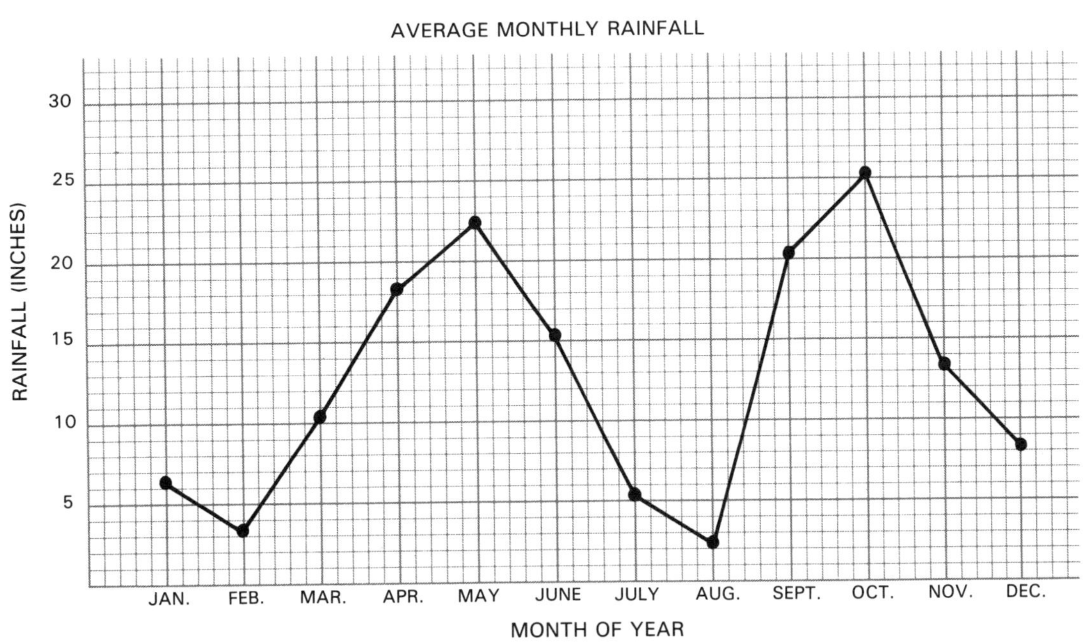

11. 15% fatalities by drugs and alcohol
    Base = 1,000,000. Portion = 150,000. Find rate.
    $R = \frac{P}{B} = \frac{150,000}{1,000,000} = 0.15 = 15\%$ for illness and disease

$$\begin{array}{r} 40 \\ 10 \\ 15 \\ +\ 20 \\ \hline 85\% \end{array} \qquad \begin{array}{r} 100 \\ -\ 85 \\ \hline 15\% \end{array}$$

12. 400,000 people
    Base = 1,000,000. Rate = 40%. Find portion.
    $P = B \times R = 1,000,000 \times 0.40 = 400,000$ people
13. 100,000 people
    Base = 1,000,000. Rate = 10%. Find portion.
    $P = B \times R = 1,000,000 \times 0.10 = 100,000$ people
14. 15%
15.

PRODUCTION OF PAPER PRODUCTS

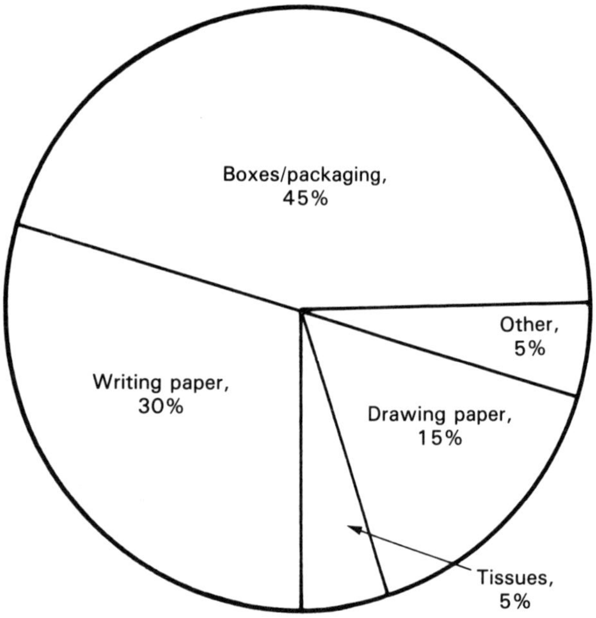

A: $P = B \times R = 360° \times 0.45 = 162°$
B: $P = B \times R = 360° \times 0.30 = 108°$
C: $P = B \times R = 360° \times 0.05 = 18°$
D: $P = B \times R = 360° \times 0.15 = 54°$

$$\begin{array}{r} 45 \\ 30 \\ 5 \\ +\ 15 \\ \hline 95\% \end{array} \qquad \begin{array}{r} 100 \\ -\ 95 \\ \hline 5\% \end{array}$$

Chapter 7
UNITS OF MEASURE

Objectives

After studying this chapter, the student will be able to:

- Explain base units and derived units.
- Convert units using a conversion bar.
- Convert units using a factor.
- Solve word problems using a conversion bar.
- Perform operations of arithmetic on denominate numbers.
- Compute travel time between time zones.

ANSWERS TO TEXT, page 88

Converting Units

1. 15,840

$$\left(\frac{3\ \text{mi.}}{}\right)\left(\frac{5280\ \text{ft.}}{1\ \text{mi.}}\right) = 15,840\ \text{ft.}$$

2. 4.1

$$\left(\frac{588\ \text{in.}^2}{}\right)\left(\frac{1\ \text{ft.}^2}{144\ \text{in.}^2}\right) = 4.083\ \text{ft.}^2$$

3. 406,080

$$\left(\frac{235\ \text{ft.}^3}{}\right)\left(\frac{1728\ \text{in.}^3}{1\ \text{ft.}^3}\right) = 406,080\ \text{in.}^3$$

4. 6

$$\left(\frac{96\ \text{oz.}}{}\right)\left(\frac{1\ \text{lb.}}{16\ \text{oz.}}\right) = 6\ \text{lb.}$$

5. 8

$$\left(\frac{32\ \text{cups}}{}\right)\left(\frac{1\ \text{qt.}}{4\ \text{cups}}\right) = 8\ \text{qt.}$$

6. 78 500
7. 650

$$\left(\frac{65\ 000\ \text{dm}^2}{}\right)\left(\frac{1\ \text{m}^2}{100\ \text{dm}^2}\right) = 650\ \text{m}^2$$

8. 8.5

$$\left(\frac{8500\ \text{m}^3}{}\right)\left(\frac{1\ \text{dam}^3}{1000\ \text{m}^3}\right) = 8.5\ \text{dam}^3$$

9. 20 000
10. 0.250
11. 8736

$$\left(\frac{52\ \text{wk.}}{}\right)\left(\frac{168\ \text{hr.}}{1\ \text{wk.}}\right) = 8736\ \text{hr.}$$

12. 86.7

$$\left(\frac{2080\ \text{hr.}}{}\right)\left(\frac{1\ \text{day}}{24\ \text{hr.}}\right) = 86.667\ \text{days}$$

13. 744

$$\left(\frac{31\ \cancel{\text{days}}}{}\right)\left(\frac{24\text{ hr.}}{1\ \cancel{\text{day}}}\right) = 744\text{ hr.}$$

14. 10,080

$$\left(\frac{1\ \cancel{\text{wk.}}}{}\right)\left(\frac{7\ \cancel{\text{day}}}{1\ \cancel{\text{wk.}}}\right)\left(\frac{1440\text{ hr.}}{1\ \cancel{\text{day}}}\right) = 10{,}080\text{ hr.}$$

15. 3.6

$$\left(\frac{12{,}900\ \cancel{\text{sec.}}}{}\right)\left(\frac{1\text{ hr.}}{3600\ \cancel{\text{sec.}}}\right) = 3.583\text{ hr.}$$

16. 58.7

$$\left(\frac{40\ \cancel{\text{mi.}}}{\cancel{\text{hr.}}}\right)\left(\frac{1\ \cancel{\text{hr.}}}{3600\text{ sec.}}\right)\left(\frac{5280\text{ ft.}}{1\ \cancel{\text{mi.}}}\right) = 58{,}667\text{ ft./sec.}$$

17. 4800

$$\left(\frac{80\text{ words}}{\cancel{\text{min.}}}\right)\left(\frac{60\ \cancel{\text{min.}}}{1\text{ hr.}}\right) = 4800\text{ words/hr.}$$

18. 0.075

$$\left(\frac{75\text{ L}}{\cancel{\text{km}}}\right)\left(\frac{1\ \cancel{\text{km}}}{1000\text{ m}}\right) = 0.075\text{ L/m}$$

19. 0.06

$$\left(\frac{60\ \cancel{\text{g}}}{\text{carton}}\right)\left(\frac{1\text{ kg.}}{1000\ \cancel{\text{g}}}\right) = 0.06\text{ kg/carton}$$

20. 0.868

$$\left(\frac{125\text{ lb.}}{\cancel{\text{ft.}^2}}\right)\left(\frac{1\ \cancel{\text{ft.}^2}}{144\text{ in.}^2}\right) = 0.868\text{ lb./in.}^2$$

21. 96.54
   $60 \times 1.609 = 96.54$ km
22. 344.32
   $32 \times 10.76 = 344.32$ ft.^{2}
23. 5.04
   $180 \times 0.028 = 5.04$ m^3
24. 189.63
   $86 \times 2.205 = 189.63$ lb.
25. 23.76
   $90 \times 0.264 = 23.76$ gal.

Arithmetic with Denominate Numbers

26. 18 yd., 1 ft., 9 in.
27. 2 yd.^{2}, 122 in.^{2}
28. 13 gal., 1 qt.
29. 1 ton, 1700 lb.
30. 49 964 g
31. 2 wk., 8 days, 18 hr., 51 min.
32. 9 yd., 2 ft., 6 in.
33. 55 gal.
34. 140 ft.^{2}
35. 1 day, 12 hr., 2 2/3 min.

Word Problems with Units of Measure

36. 2.0625 tons

$$\left(\frac{4125\ \cancel{\text{lb.}}}{}\right)\left(\frac{1\text{ ton}}{2000\ \cancel{\text{lb.}}}\right) = 2.0625\text{ tons}$$

37. 6.1 mi.

$$\left(\frac{32{,}000\ \cancel{\text{ft.}}}{}\right)\left(\frac{1\text{ mi.}}{5280\ \cancel{\text{ft.}}}\right) = 6.0606\text{ mi.}$$

38. 221.4 m
   $726 \times 0.305 = 221.43$ m
39. 96 000 000 m^3
   $125{,}600{,}000 \times 0.7646 = 96\ 033\ 760$ m^3
40. a = 22.875 cm, b = 30.5 cm, c = 15.25 cm
   a = 2 − 1.25 = 0.75 in. ⇒ 0.75 ft.
      $0.75 \times 0.305 \times 100 = 22.875$ cm
   b = 2.5 − 1.5 = 1 in. ⇒ 1 ft.
      $1 \times 0.305 \times 100 = 30.5$ cm
   c = 1.5 − (0.5 + 0.5) = 0.5 in. ⇒ 0.5 ft.
      $0.5 \times 0.305 \times 100 = 15.25$ cm
41. 62.1 mph
   $100 \times 0.621 = 62.1$ mi./hr.
42. 9.76 m/s^2
   $32 \times 0.305 = 9.76$ m/s^2
43. 1034.4 g/cm^2
   $14.7 \times 0.454 \times 1000 = 6673.8$ g/in.^{2}

$$\frac{6673.8\text{ g}}{1 \times 6.452} = \frac{6673.8\text{ g}}{6.452\text{ cm}^2} = 1034.377\text{ g/cm}^2$$

44. (a) 3.03 mi. long, (b) 18.2 mph

$$\left(\frac{40\ \cancel{\text{ft.}}}{\cancel{\text{car}}}\right)\left(\frac{40\ \cancel{\text{car}}}{\cancel{\text{min.}}}\right)\left(10\ \cancel{\text{min.}}\right)\left(\frac{1\text{ mi.}}{5280\ \cancel{\text{ft.}}}\right) = 3.030\text{ mi.}$$

$$\left(\frac{3.03\text{ mi.}}{10\ \cancel{\text{min.}}}\right)\left(\frac{60\ \cancel{\text{min.}}}{1\text{ hr.}}\right) = 18.182\text{ mi./hr.}$$

45. $10,200 labor cost

$$\left(\frac{200\ \cancel{\text{yd.}^3}}{}\right)\left(\frac{3\ \cancel{\text{hr.}}}{\cancel{\text{yd.}^3}}\right)\left(\frac{\$17}{\cancel{\text{hr.}}}\right) = \$10{,}200$$

Times Zones and Travel Time

46. 12:15 a.m.
47. 5:08 p.m.
48. 1731 hours
49. 5 hr., 20 min.
50. 12:20 a.m.

ANSWERS TO WORKBOOK, page 55

Converting Units

1. 12, 4

$$\left(\frac{144\ \cancel{\text{in.}}}{}\right)\left(\frac{1\text{ ft.}}{12\ \cancel{\text{in.}}}\right) = 12\text{ ft.}$$

$$\left(\frac{12\ \cancel{\text{ft.}}}{}\right)\left(\frac{1\text{ yd.}}{3\ \cancel{\text{ft.}}}\right) = 4\text{ yd.}$$

2. 3, 108

$$\left(\frac{9\ \cancel{\text{ft.}}}{}\right)\left(\frac{1\text{ yd.}}{3\ \cancel{\text{ft.}}}\right) = 3\text{ yd.}$$

$$\left(\frac{3\ \cancel{\text{yd.}}}{}\right)\left(\frac{36\text{ in.}}{1\ \cancel{\text{yd.}}}\right) = 108\text{ in.}$$

3. 21,120, 7040

$$\left(\frac{4\ \cancel{\text{mi.}}}{}\right)\left(\frac{5280\text{ ft.}}{1\ \cancel{\text{mi.}}}\right) = 21{,}120\text{ ft.}$$

$$\left(\frac{21{,}120\ \cancel{\text{ft.}}}{}\right)\left(\frac{1\text{ yd.}}{3\ \cancel{\text{ft.}}}\right) = 7040\text{ yd.}$$

4. 36, 5184

$$\left(\frac{4\text{ yd.}^2}{}\right)\left(\frac{9\text{ ft.}^2}{1\text{ yd.}^2}\right) = 36\text{ ft.}^2$$

$$\left(\frac{4\text{ yd.}^2}{}\right)\left(\frac{1296\text{ in.}^2}{1\text{ yd.}^2}\right) = 5184\text{ in.}^2$$

5. 1, 1296
6. 279,936, 6

$$\left(\frac{162\text{ ft.}^3}{}\right)\left(\frac{1728\text{ in.}^3}{1\text{ ft.}^3}\right) = 279{,}936\text{ in.}^3$$

$$\left(\frac{162\text{ ft.}^3}{}\right)\left(\frac{1\text{ yd.}^3}{27\text{ ft.}^3}\right) = 6\text{ yd.}^3$$

7. 1840

$$\left(\frac{115\text{ lb.}}{}\right)\left(\frac{16\text{ oz.}}{1\text{ lb.}}\right) = 1840\text{ oz.}$$

8. 16

$$\left(\frac{256\text{ oz.}}{}\right)\left(\frac{1\text{ lb.}}{16\text{ oz.}}\right) = 16\text{ lb.}$$

9. 4.25, 2.125

$$\left(\frac{68\text{ fl. oz.}}{}\right)\left(\frac{1\text{ pt.}}{16\text{ fl. oz.}}\right) = 4.25\text{ pt.}$$

$$\left(\frac{68\text{ ft. oz.}}{}\right)\left(\frac{1\text{ qt.}}{32\text{ fl. oz.}}\right) = 2.125\text{ qt.}$$

10. 220, 27.5

$$\left(\frac{110\text{ qt.}}{}\right)\left(\frac{2\text{ pt.}}{1\text{ qt.}}\right) = 220\text{ pt.}$$

$$\left(\frac{110\text{ qt.}}{}\right)\left(\frac{1\text{ gal.}}{4\text{ qt.}}\right) = 27.5\text{ gal.}$$

11. 2, 200
12. 3, 300
13. 3.5, 35
14. 600, 0.6
15. 400, 40 000

$$\left(\frac{4\text{ m}^2}{}\right)\left(\frac{100\text{ dm}^2}{1\text{ m}^2}\right) = 400\text{ dm}^2$$

$$\left(\frac{4\text{ m}^2}{}\right)\left(\frac{10\ 000\text{ cm}^2}{1\text{ m}^2}\right) = 40\ 000\text{ cm}^2$$

16. 5, 5 000 000

$$\left(\frac{5000\text{ m}^3}{}\right)\left(\frac{1\text{ dam}^3}{1000\text{ m}^3}\right) = 5\text{ dam}^3$$

$$\left(\frac{5000\text{ m}^3}{}\right)\left(\frac{1000\text{ dm}^3}{1\text{ m}^3}\right) = 5\ 000\ 000\text{ dm}^3$$

17. 2000, 20
18. 500, 0.5
19. 650, 0.65
20. 45, 4.5

21. 2, 23.9

$$\left(\frac{104\text{ wk.}}{}\right)\left(\frac{1\text{ yr.}}{52.18\text{ wk.}}\right) = 1.993\text{ yr.}$$

$$\left(\frac{104\text{ wk.}}{}\right)\left(\frac{1\text{ mo.}}{4.35\text{ wk.}}\right) = 23.908\text{ mo.}$$

22. 260, 59.8

$$\left(\frac{1820\text{ days}}{}\right)\left(\frac{1\text{ wk.}}{7\text{ days}}\right) = 260\text{ wk.}$$

$$\left(\frac{1820\text{ days}}{}\right)\left(\frac{1\text{ mo.}}{30.44\text{ days}}\right) = 59.790\text{ mo.}$$

23. 252, 6048

$$\left(\frac{36\text{ wk.}}{}\right)\left(\frac{7\text{ days}}{1\text{ wk.}}\right) = 252\text{ days}$$

$$\left(\frac{36\text{ wk.}}{}\right)\left(\frac{168\text{ hr.}}{1\text{ wk.}}\right) = 6048\text{ hr.}$$

24. 3, 504

$$\left(\frac{21\text{ days}}{}\right)\left(\frac{1\text{ wk.}}{7\text{ days}}\right) = 3\text{ wk.}$$

$$\left(\frac{21\text{ days}}{}\right)\left(\frac{24\text{ hr.}}{1\text{ day}}\right) = 504\text{ hr.}$$

25. 120, 7200

$$\left(\frac{5\text{ days}}{}\right)\left(\frac{24\text{ hr.}}{1\text{ day}}\right) = 120\text{ hr.}$$

$$\left(\frac{5\text{ days}}{}\right)\left(\frac{1440\text{ min.}}{1\text{ day}}\right) = 7200\text{ min.}$$

26. 8, 192

$$\left(\frac{11{,}520\text{ min.}}{}\right)\left(\frac{1\text{ day}}{1440\text{ min.}}\right) = 8\text{ days}$$

$$\left(\frac{11{,}520\text{ min.}}{}\right)\left(\frac{1\text{ hr.}}{60\text{ min.}}\right) = 192\text{ hr.}$$

27. 60, 3600

$$\left(\frac{2.5\text{ days}}{}\right)\left(\frac{24\text{ hr.}}{1\text{ day}}\right) = 60\text{ hr.}$$

$$\left(\frac{2.5\text{ days}}{}\right)\left(\frac{1440\text{ min.}}{1\text{ day}}\right) = 3600\text{ min.}$$

28. 8784, 527,040

$$\left(\frac{1\text{ leap year}}{}\right)\left(\frac{366\text{ days}}{1\text{ leap year}}\right)\left(\frac{24\text{ hr.}}{1\text{ day}}\right) = 8784\text{ hr.}$$

$$\left(\frac{8784\text{ hr.}}{}\right)\left(\frac{60\text{ sec.}}{1\text{ hr.}}\right) = 527{,}040\text{ min.}$$

29. 90, 1.5

$$\left(\frac{5400\text{ sec.}}{}\right)\left(\frac{1\text{ min.}}{60\text{ sec.}}\right) = 90\text{ min.}$$

$$\left(\frac{5400\text{ sec.}}{}\right)\left(\frac{1\text{ hr.}}{3600\text{ sec.}}\right) = 1.5\text{ hr.}$$

30. 720, 4.3

$$\left(\frac{30\ \cancel{\text{days}}}{}\right)\left(\frac{24\ \text{hr.}}{1\ \cancel{\text{day}}}\right) = 720\ \text{hr.}$$

$$\left(\frac{30\ \cancel{\text{days}}}{}\right)\left(\frac{1\ \text{wk.}}{7\ \cancel{\text{days}}}\right) = 4.286\ \text{wk.}$$

31. 0.0167, 88

$$\left(\frac{60\ \text{mi.}}{\cancel{\text{hr.}}}\right)\left(\frac{1\ \cancel{\text{hr.}}}{3600\ \text{sec.}}\right) = 0.0167\ \text{mi./sec.}$$

$$\left(\frac{60\ \cancel{\text{mi.}}}{\cancel{\text{hr.}}}\right)\left(\frac{5280\ \text{ft.}}{1\ \cancel{\text{mi.}}}\right)\left(\frac{1\ \cancel{\text{hr.}}}{3600\ \text{sec.}}\right) = 88\ \text{ft./sec.}$$

32. 80,000, 1333.3

$$\left(\frac{80{,}000\ \text{m}}{\cancel{\text{h}}}\right)\left(\frac{1\ \cancel{\text{h}}}{60\ \text{s}}\right) = 1333.33\ \text{m/s}$$

33. 5, 0.05

$$\left(\frac{20\ \text{mi.}}{\cancel{\text{gal.}}}\right)\left(\frac{1\ \cancel{\text{gal.}}}{4\ \text{qt.}}\right) = 5\ \text{mi./qt.}$$

34. 2, 7200

$$\left(\frac{120\ \text{words}}{\cancel{\text{min.}}}\right)\left(\frac{1\ \cancel{\text{min.}}}{60\ \text{sec.}}\right) = 2\ \text{words/sec.}$$

$$\left(\frac{120\ \text{words}}{\cancel{\text{min.}}}\right)\left(\frac{60\ \cancel{\text{min.}}}{1\ \text{hr.}}\right) = 7200\ \text{words/hr.}$$

35. 1440, 1

$$\left(\frac{60\ \text{gal.}}{\cancel{\text{hr.}}}\right)\left(\frac{24\ \cancel{\text{hr.}}}{1\ \text{day}}\right) = 1440\ \text{gal./day}$$

$$\left(\frac{60\ \text{gal.}}{\cancel{\text{hr.}}}\right)\left(\frac{1\ \cancel{\text{hr.}}}{60\ \text{min.}}\right) = 1\ \text{gal./min.}$$

36. 0.075, 0.075

$$\left(\frac{75\ \text{L}}{\cancel{\text{km}}}\right)\left(\frac{1\ \cancel{\text{km}}}{1000\ \text{m}}\right) = 0.075\ \text{L/m}$$

37. 400, 2.8

$$\left(\frac{25\ \cancel{\text{lb.}}}{\text{ft.}^2}\right)\left(\frac{16\ \text{oz.}}{1\ \cancel{\text{lb.}}}\right) = 400\ \text{oz./ft.}^2$$

$$\left(\frac{400\ \text{oz.}}{\cancel{\text{ft.}^2}}\right)\left(\frac{1\ \cancel{\text{ft.}^2}}{144\ \text{in.}^2}\right) = 2.778\ \text{oz./in.}^2$$

38. 0.21, 750

$$\left(\frac{1100\ \cancel{\text{ft.}}}{\text{sec.}}\right)\left(\frac{1\ \text{mi.}}{5280\ \cancel{\text{ft.}}}\right) = 0.208\ \text{mi./sec.}$$

$$\left(\frac{1100\ \cancel{\text{ft./sec.}}}{}\right)\left(\frac{1\ \text{mi./hr.}}{1.467\ \cancel{\text{ft./sec.}}}\right) = 749.83\ \text{mi./hr.}$$

39. 21.8, 1920

$$\left(\frac{32\ \cancel{\text{ft./sec.}}}{}\right)\left(\frac{1\ \text{mi./hr.}}{1.467\ \cancel{\text{ft./sec.}}}\right) = 21.813\ \text{mi./hr.}$$

$$\left(\frac{32\ \text{ft.}}{\cancel{\text{sec.}}}\right)\left(\frac{60\ \cancel{\text{sec.}}}{1\ \text{min.}}\right) = 1920\ \text{ft./min.}$$

40. 33.4, 9.3

$$\left(\frac{556\ \cancel{\text{g}}}{\cancel{\text{min.}}}\right)\left(\frac{1\ \text{kg}}{1000\ \cancel{\text{g}}}\right)\left(\frac{60\ \cancel{\text{min.}}}{1\ \text{h}}\right) = 33.36\ \text{kg/h}$$

$$\left(\frac{556\ \text{g}}{\cancel{\text{min.}}}\right)\left(\frac{1\ \cancel{\text{min.}}}{60\ \text{s}}\right) = 9.267\ \text{g/s}$$

41. 51.488, 51,488
$32 \times 1.609 = 51.488$ km

42. 109.4, 328.1
$100 \times 1.094 = 109.4$ yd.
$100 \times 3.281 = 328.1$ ft.

43. 2.364, 0.197
$6 \times 0.394 = 2.364$ in.
$0.06 \times 3.281 = 0.197$ ft.

44. 1.83, 183
$6 \times 0.305 = 1.83$ m

45. 2.418, 24,180
$26 \times 0.093 = 2.418\ \text{m}^2$

$$\left(\frac{2.418\ \cancel{\text{m}^2}}{}\right)\left(\frac{10\,000\ \text{cm}^2}{1\ \cancel{\text{m}^2}}\right) = 24\,180\ \text{cm}^2$$

46. 344.3, 49,582
$32 \times 10.76 = 344.32\ \text{ft.}^2$

$$\left(\frac{344.32\ \cancel{\text{ft.}^2}}{}\right)\left(\frac{144\ \text{in.}^2}{1\ \cancel{\text{ft.}^2}}\right) = 49{,}582.08\ \text{in.}^2$$

47. 15 876, 15.876
$560 \times 28.35 = 15\,876$ g

48. 49.94, 49 940
$110 \times 0.454 = 49.94$ kg

49. 119.1, 1905.1
$54 \times 2.205 = 119.07$ lb.

$$\left(\frac{119.07\ \cancel{\text{lb.}}}{}\right)\left(\frac{16\ \text{oz.}}{1\ \cancel{\text{lb.}}}\right) = 1905.12\ \text{oz.}$$

50. 20, 79.9
$75.7 \times 0.264 = 19.9848$ gal.
$75.7 \times 1.056 = 79.9392$ qt.

Arithmetic with Denominate Numbers

51. 20 ft.^{2}
52. 12 yd.
53. 20 gal., 2 qt.
54. 3 wk., 2 days, 10 hr., 16 min., 19 sec.
55. 20 lb., 12 oz.
56. 6 hr., 58 min.
57. 28 lb., 4 oz.
58. 6 gal., 1 pt.
59. 3 yd., 2 ft., 6 in.
60. 2 wk., 4 days, 12 hr., 21 min.
61. 10 yd., 1 ft., 6 in.
62. 28 ft.^{2}
63. 720 m^3
64. 203 lb.
65. $205, 60¢
66. 1 hr., 15 min.
67. 2 yd., 11 2/3 in.

68. 3 gal., 1 1/2 pt.
69. 3 ft., 7 in.
70. 1 1/2 pt.

Time Zones and Travel Time

71. 0220 hours
72. 1840 hours
73. 2:50 p.m.
74. 9:05 a.m.
75. (a) 1:30 a.m., (b) 2:30 a.m., (c) 3:30 a.m.
76. (a) 9:00 p.m., (b) 11:00 p.m., (c) 12:00 p.m.
77. (a) 12:45 a.m., (b) 1:45 a.m., (c) 3:45 a.m.
78. (a) 2:30 p.m., (b) 3:30 p.m., (c) 4:30 p.m.
79. 3 days, 17 hr., 45 min.
80. 2 hr., 15 min.

ANSWERS TO TEST MASTERS, page 106

1. 2.44
2. 2.47
3. 8
4. 96
5. 5
6. 6000
7. 20
8. 20 000 000
9. 20,160
10. 4
11. 102.7
12. 10
13. 0.12
14. 0.48
15. 21.8
16. 96.54
17. 68.1
18. 26.4
19. 38.1
20. 484.2
21. 61 gal., 1 qt.
22. 49 yd., 1 ft., 6 in.
23. 1500 lb.
24. 9 gal., 1 qt.
25. 5 d., 4 hr., 52 min.
26. 3:15 a.m.
27. 12:50 a.m.
28. 2300 hours
29. 3:30 p.m.
30. 2:00 a.m.
31. 12:40 p.m.
32. 8:00 a.m., Tuesday
33. 5 hr., 35 min.

Chapter 8
PERIMETER, AREA, AND VOLUME

Objectives

After studying this chapter, the student will be able to:

- Explain basic terminology of geometry.
- Compute perimeter and area of plane shapes.
- Compute volume and surface area of solids.

ANSWERS TO TEXT, page 109
Perimeter

1. 180 ft.
   Formula 8-2: $p = 2l + 2w$
   Substitution: $p = 2(30 \text{ ft.}) + 2(60 \text{ ft.})$
   Solution: $p = 180 \text{ ft.}$
2. 84 in.
   Formula 8-2: $p = 2l + 2w$
   Substitution: $p = 2(24 \text{ in.}) + 2(18 \text{ in.})$
   Solution: $p = 84 \text{ in.}$
3. 70 ft.
   Conversion: 120 in. = 10 ft.
   Formula 8-2: $p = 2l + 2w$
   Substitution: $p = 2(10 \text{ ft.}) + 2(25 \text{ ft.})$
   Solution: $p = 70 \text{ ft.}$
4. 1650 ft. (550 yd.)
   Conversion: 150 yd. = 450 ft.
   Formula 8-2: $p = 2l + 2w$
   Substitution: $p = 2(450 \text{ ft.}) + 2(375 \text{ ft.})$
   Solution: $p = 1650 \text{ ft.} = 550 \text{ yd.}$
5. 52 in.
   Outside of frame = 12 in. × 14 in.
   Formula 8-2: $p = 2l + 2w$
   Substitution: $p = 2(12 \text{ in.}) + 2(14 \text{ in.})$
   Solution: $p = 52 \text{ in.}$
6. 900 ft., 300 ft., 300 ft.
   Formula 8-2: $p = 2l + 2w$
   Substitution: $2400 \text{ ft.} = 2(900 \text{ ft.}) + 2w$
   Transposition: $w = \dfrac{2400 \text{ ft.} - 2\ (900 \text{ ft.})}{2}$
   Solution: $w = 300 \text{ ft.}$
7. 60 in.
   Formula 8-3: $p = 4s$
   Substitution: $p = 4(15 \text{ in.})$
   Solution: $p = 60 \text{ in.}$
8. 304 yd.
   Formula 8-3: $p = 4s$
   Substitution: $p = 4(76 \text{ yd.})$
   Solution: $p = 304 \text{ yd.}$
9. 20 m
   Formula 8-3: $p = 4s$
   Substitution: $80 \text{ m} = 4s$
   Transposition: $s = \dfrac{80 \text{ m}}{4}$
   Solution: $s = 20 \text{ m}$
10. 324 in. (27 ft.)
    Conversion: 6 ft., 9 in. = 81 in.
    Formula 8-3: $p = 4s$
    Substitution: $p = 4(81 \text{ in.})$
    Solution: $p = 324 \text{ in.} = 27 \text{ ft.}$
11. 18.67 ft. (18 ft., 8 in.)
    Formula 8-3: $p = 4s$
    Substitution: $p = 4(56 \text{ in.})$
    Solution: $p = 224 \text{ in.} = 18.67 \text{ ft.}$
    $= 18 \text{ ft.}, 8 \text{ in.}$
12. 3 m
    Formula 8-3: $p = 4s$
    Substitution: $p = 4(75 \text{ cm})$
    Solution: $p = 300 \text{ cm} = 3 \text{ m}$
13. 52 in.
    Formula 8-4: $p = \text{side a} + \text{side b} + \text{side c}$
    Substitution: $p = 14 \text{ in.} + 20 \text{ in.} + 18 \text{ in.}$
    Solution: $p = 52 \text{ in.}$

14. 30 ft.
Formula 8-4: p = side a + side b + side c
Substitution: 86 ft. = 34 ft. + 22 ft. + side c
Transposition: side c = 86 ft. − 34 ft. − 22 ft.
Solution: side c = 30 ft.

15. 75.40 in.
Formula 8-6: $p = 2\pi r$
Substitution: $p = 2 \times \pi \times 12$ in.
Solution: p = 75.40 in.

16. 471.24 cm
Formula 8-6: $p = 2\pi r$
Substitution: $p = 2 \times \pi \times 75$ cm
Solution: p = 471.24 cm

17. 56.55 in.
Formula 8-7: $p = \pi d$
Substitution: $p = \pi \times 18$ in.
Solution: p = 56.55 in.

18. 2.86 in.
Formula 8-6: $p = 2\pi r$
Substitution: 18 in. $= 2 \times \pi \times r$
Transposition: $r = \frac{18 \text{ in.}}{2\pi}$
Solution: r = 2.86 in.

19. 9.55 cm
Formula 8-7: $p = \pi d$
Substitution: 30 cm $= \pi \times d$
Transposition: $d = \frac{30 \text{ cm}}{\pi}$
Solution: d = 9.55 cm

20. 48 m
Formula 8-1: p = side a + side b + side c + . . .
Substitution: p = (9 + 2 + 6 + 1 + 1 + 1 + 1 + 1 + 6 + 3 + 5 + 3 + 2 + 3 + 2 + 2) m
Solution: p = 48 m

Area

21. 360 in.²
Formula 8-8: $A = l \times w$
Substitution: A = 18 in. × 20 in.
Solution: A = 360 in.²

22. 360 tiles
Formula 8-8: $A = l \times w$
Substitution: A = 24 ft. × 15 ft.
Area of floor: A = 360 ft.²
Conversion: 12 in. = 1 ft.
Formula 8-9: $A = s^2$
Substitution: A = 1 ft. × 1 ft.
Area of tile: A = 1 ft.²
Solution: $\left(\frac{360 \text{ ft.}^2}{}\right)\left(\frac{1 \text{ tile}}{1 \text{ ft.}^2}\right) = 360$ tiles

23. 70 tiles
Formula 8-8: $A = l \times w$
Substitution: A = 14 ft. × 20 ft.
Area of ceiling: A = 280 ft.²
Formula 8-9: $A = s^2$
Substitution: A = 2 ft. × 2 ft.
Area of tile: A = 4 ft.²
Solution: $\left(\frac{280 \text{ ft.}^2}{}\right)\left(\frac{1 \text{ tile}}{4 \text{ ft.}^2}\right) = 70$ tiles

24. 21.33 yd.²
Formula 8-8: $A = l \times w$
Substitution: A = 16 ft. × 12 ft.
Solution: A = 192 ft.² = 21.33 yd.²

25. 2 rolls
Walls:
Formula 8-8: $A = l \times w$
Substitution: A = 12 ft. × 8 ft.
Area of 1 wall: A = 96 ft.²
Area of 4 walls: A = 96 ft.² × 4 = 384 ft.²
Wallpaper:
Conversion: 24 in. = 2 ft.
Formula 8-8: $A = l \times w$
Substitution: A = 100 ft. × 2 ft.
Area of wallpaper: A = 200 ft.²
Rolls of paper:

$$\left(\frac{384 \text{ ft.}^2}{}\right)\left(\frac{1 \text{ roll}}{200 \text{ ft.}^2}\right) = 1.92 \text{ rolls}$$

Solution: Buy 2 rolls.

26. 3 rolls
10-foot walls:
Formula 8-8: $A = l \times w$
Substitution: A = 10 ft. × 8 ft.
Area of 1 wall: A = 80 ft.²
Area of 2 walls: A = 80 ft.² × 2 = 160 ft.²

16-foot walls:
Formula 8-8: $A = l \times w$
Substitution: A = 16 ft. × 8 ft.
Area of 1 wall: A = 128 ft.²
Area of 2 walls: A = 128 ft.² × 2 = 256 ft.²

All walls:
Area of all walls: A = 256 ft.² + 160 ft.² = 416 ft.²

Wallpaper:
Conversion: 18 in. = 1.5 ft.
Formula 8-8: $A = l \times w$
Substitution: A = 100 ft. × 1.5 ft.
Area of wallpaper: A = 150 ft.²
Rolls of paper:

$$\left(\frac{416 \text{ ft.}^2}{}\right)\left(\frac{1 \text{ roll}}{150 \text{ ft.}^2}\right) = 2.77 \text{ rolls}$$

Solution: Buy 3 rolls.

27. 4 gal.
60-foot walls:
Formula 8-8: $A = l \times w$
Substitution: A = 60 ft. × 8 ft.
Area of 1 wall: A = 480 ft.²
Area of 2 walls: A = 480 ft.² × 2 = 960 ft.²

32-foot walls:
Formula 8-8: $A = l \times w$
Substitution: A = 32 ft. × 8 ft.
Area of 1 wall: A = 256 ft.²
Area of 2 walls: A = 256 ft.² × 2 = 512 ft.²

All walls:
Area of all walls: A = 960 ft.² + 512 ft.² = 1472 ft.²

Gallons of paint needed:

$$\left(\frac{1472 \text{ ft.}^2}{}\right)\left(\frac{1 \text{ gal.}}{400 \text{ ft.}^2}\right) = 3.68 \text{ gal.}$$

Solution: Buy 4 gal.

28. 225 in.²
Formula 8-9: $A = s^2$

Substitution: $A = 15 \text{ in.} \times 15 \text{ in.}$
Solution: $A = 225 \text{ in.}^2$

29. 11.31 cm
Formula 8-9: $A = s^2$
Substitution: $128 \text{ cm}^2 = s^2$
Transposition: $s = \sqrt{128 \text{ cm}^2}$
Solution: $s = 11.31 \text{ cm}$

30. 175 ft.²
Formula 8-10: $A = 1/2\ (b \times h)$
Substitution: $A = 1/2 \times 25 \text{ ft.} \times 14 \text{ ft.}$
Solution: $A = 175 \text{ ft.}^2$

31. 128 ft.²
Formula 8-10: $A = 1/2\ (b \times h)$
Substitution: $A = 1/2 \times 32 \text{ ft.} \times 8 \text{ ft.}$
Solution: $A = 128 \text{ ft.}^2$

32. 364 ft.²
Triangular portion:
Height: 18 ft. − 8 ft. = 10 ft.
Formula 8-10: $A = 1/2\ (b \times h)$
Substitution: $A = 1/2 \times 28 \text{ ft.} \times 10 \text{ ft.}$
Area of triangular portion: $A = 140 \text{ ft.}^2$

Rectangular portion:
Formula 8-8: $A = l \times w$
Substitution: $A = 28 \text{ ft.} \times 8 \text{ ft.}$
Area of rectangular portion: $A = 224 \text{ ft.}^2$
Solution: $A = 140 \text{ ft.}^2 + 224 \text{ ft.}^2 = 364 \text{ ft.}^2$

33. 706.86 in.²
Formula 8-11: $A = \pi r^2$
Substitution: $A = \pi \times (15 \text{ in.})^2$
Solution: $A = 706.86 \text{ in.}^2$

34. 615.75 m²
Formula 8-12: $A = \pi d^2/4$
Substitution: $A = \pi \times (28 \text{ m})^2/4$
Solution: $A = 615.75 \text{ m}^2$

Volume

35. 1344 in.³
Formula 8-13: $V = l \times w \times h$
Substitution: $V = 14 \text{ in.} \times 12 \text{ in.} \times 8 \text{ in.}$
Solution: $V = 1344 \text{ in.}^3$

36. 1.50 ft.³
Formula 8-13: $V = l \times w \times h$
Substitution: $V = 1.5 \text{ ft.} \times 2 \text{ ft.} \times 0.5 \text{ ft.}$
Solution: $V = 1.50 \text{ ft.}^3$

37. 1792 ft.³
Formula 8-13: $V = l \times w \times h$
Substitution: $V = 14 \text{ ft.} \times 16 \text{ ft.} \times 8 \text{ ft.}$
Solution: $V = 1792 \text{ ft.}^3$

38. 46 656 cm³
Formula 8-14: $V = s^3$
Substitution: $V = (36 \text{ cm})^3$
$= 36 \text{ cm} \times 36 \text{ cm} \times 36 \text{ cm}$
Solution: $V = 46\ 656 \text{ cm}^3$

39. 308,831.12 in.³
Formula 8-18: $V = \pi r^2 \times h$
Substitution: $V = \pi \times (32 \text{ in.})^2 \times 96 \text{ in.}$
Solution: $V = 308{,}831.12 \text{ in.}^3$

40. 29,080.03 gal.
Formula 8-13: $V = l \times w \times h$
Substitution: $V = 36 \text{ ft.} \times 12 \text{ ft.} \times 9 \text{ ft.}$
Solution: $V = 3888 \text{ ft.}^3 = 29{,}080.03 \text{ gal.}$

ANSWERS TO WORKBOOK, page 69

Perimeter

1. 231 5/6 m (231.83 m)
Formula 8-2: $\mathbf{p = 2l + 2w}$
Substitution: $\mathbf{p = 2(35\ 1/4 \text{ m}) + 2(80\ 2/3 \text{ m})}$
Solution: $\mathbf{p = 231\ 5/6 \text{ m}}$

2. 41 2/3 yd. (41.67 yd.)
Formula 8-2: $\mathbf{p = 2l + 2w}$
Substitution: $\mathbf{p = 2(5\ 1/3 \text{ yd.}) + 2(15\ 1/2 \text{ yd.})}$
Solution: $\mathbf{p = 41\ 2/3 \text{ yd.}}$

3. 27 m
Conversion (to meters): **820 cm = 8.2 m**
Formula 8-2: $\mathbf{p = 2l + 2w}$
Substitution: $\mathbf{p = 2(5.3 \text{ m}) + 2(8.2 \text{ m})}$
Solution: $\mathbf{p = 27 \text{ m}}$

4. 46.67 ft. (46 ft., 8 in.)
Formula 8-2: $\mathbf{p = 2l + 2w}$
Substitution: $\mathbf{p = 2(202 \text{ in.}) + 2(96 \text{ in.})}$
Perimeter of room: **p = 596 in.**
Perimeter less doorway: 596 in. − 36 in. = **560 in.**
Solution: Baseboard = 560 in. = **46.67 ft.**
= 46 ft., 8 in.

5. 12.67 ft. (12 ft., 8 in.)
Formula 8-2: $\mathbf{p = 2l + 2w}$
Substitution: $\mathbf{p = 2(42 \text{ in.}) + 2(34 \text{ in.})}$
Solution: **p** = 152 in. = **12.67 ft.**
= 12 ft., 8 in.

6. 85 in., 255 in., 255 in.
Formula 8-2: $\mathbf{p = 2l + 2w}$
Substitution: $\mathbf{680 \text{ in.} = 2(85 \text{ in.}) + 2w}$
Transposition: $\mathbf{w = \dfrac{680 \text{ in.} - 2(85 \text{ in.})}{2}}$
Solution: $\mathbf{w = 255 \text{ in.}}$

7. 200 cm
Formula 8-3: $\mathbf{p = 4s}$
Substitution: $\mathbf{p = 4(50 \text{ cm})}$
Solution: $\mathbf{p = 200 \text{ cm}}$

8. 60 m
Formula 8-3: $\mathbf{p = 4s}$
Substitution: $\mathbf{p = 4(15 \text{ m})}$
Solution: $\mathbf{p = 60 \text{ m}}$

9. 190 ft.
Formula 8-3: $\mathbf{p = 4s}$
Substitution: $\mathbf{760 \text{ ft.} = 4s}$
Transposition: $\mathbf{s = \dfrac{760 \text{ ft.}}{4}}$
Solution: $\mathbf{s = 190 \text{ ft.}}$

10. 416 in. (34 ft., 8 in.)
Conversion (to inches): **8 ft., 8 in. = 104 in.**
Formula 8-3: $\mathbf{p = 4s}$
Substitution: $\mathbf{p = 4(104 \text{ in.})}$
Solution: **p = 416 in. = 34 ft., 8 in.**

11. 14.22 yd. (14 yd., 8 in.)
Formula 8-3: $\mathbf{p = 4s}$
Substitution: $\mathbf{p = 4(128 \text{ in.})}$
Solution: **p** = 512 in. = **14.22 yd.**
= 14 yd., 8 in.

12. 276 in.
Conversion (to inches): **1 yd., 2 ft., 9 in. = 69 in.**
Formula 8-3: $\mathbf{p = 4s}$

Substitution: **p = 4(69 in.)**
Solution: **p = 276 in.**

13. 66 cm
Formula 8-4: **p = side a + side b + side c**
Substitution: **p = 32 cm + 18 cm + 16 cm**
Solution: **p = 66 cm**

14. 13 yd.
Formula 8-4: **p = side a + side b + side c**
Substitution: **29 yd. = 10 yd. + 6 yd. + side c**
Transposition: **side c = 29 yd. − 10 yd. − 6 yd.**
Solution: **side c = 13 yd.**

15. 100.53 ft.
Formula 8-6: **p = 2πr**
Substitution: **p = 2 × π × 16 ft.**
Solution: **p = 100.53 ft.**

16. 163.36 m
Formula 8-6: **p = 2πr**
Substitution: **p = 2 × π × 26 m**
Solution: **p = 163.36 m**

17. 103.67 yd.
Formula 8-7: **p = πd**
Substitution: **p = π × 33 yd.**
Solution: **p = 103.67 yd.**

18. 8.91 ft.
Formula 8-6: **p = 2πr**
Substitution: **56 ft. = 2 × π × r**
Transposition: $\mathbf{r = \dfrac{56\ ft.}{2 \times \pi}}$
Solution: **p = 8.91 ft.**

19. 47.75 m
Formula 8-7: **p = πd**
Substitution: **150 m = π × d**
Transposition: $\mathbf{d = \dfrac{150\ m}{\pi}}$
Solution: **d = 47.75 m**

20. 56 ft.
Formula 8-1: **p = side a + side b + side c + . . .**
Substitution: **p = (8 + 4 + 2 + 4 + 6 + 6 + 12 + 14) ft.**
Solution: **p = 56 ft.**

Area

21. 1344 m²
Formula 8-8: **A = *l* × w**
Substitution: **A = 56 m × 24 m**
Solution: **A = 1344 m²**

22. 96 tiles
Floor:
Formula 8-8: **A = *l* × w**
Substitution: **A = 8 ft. × 12 ft.**
Area of floor: **A = 96 ft.²**

Tile:
Conversion (to feet): **12 in. = 1 ft.**
Formula 8-8: **A = *l* × w**
Substitution: **A = 1 ft. × 1 ft.**
Area of 1 tile: **A = 1 ft.²**
Solution: $\left(\dfrac{\mathbf{96\ \cancel{ft.^2}}}{}\right)\left(\dfrac{\mathbf{1\ tile}}{\mathbf{1\ \cancel{ft.^2}}}\right) = \mathbf{96\ tiles}$

23. 40 tiles
Ceiling:
Formula 8-8: **A = *l* × w**
Substitution: **A = 10 ft. × 16 ft.**
Area of ceiling: **A = 160 ft.²**

Tile:
Formula 8-8: **A = *l* × w**
Substitution: **A = 2 ft. × 2 ft.**
Area of 1 tile: **A = 4 ft.²**
Solution: $\left(\dfrac{\mathbf{160\ \cancel{ft.^2}}}{}\right)\left(\dfrac{\mathbf{1\ tile}}{\mathbf{4\ \cancel{ft.^2}}}\right) = \mathbf{40\ tiles}$

24. 12.44 yd.²
Formula 8-8: **A = *l* × w**
Substitution: **A = 8 ft. × 14 ft.**
Solution: **A =** 112 ft.² = **12.44 yd.²**

25. 3 rolls
Walls:
Formula 8-8: **A = *l* × w**
Substitution: **A = 7.5 ft. × 14 ft.**
Area of 1 wall: **A = 105 ft.²**
Area of 4 walls: **A =** 105 ft.² × 4 = **420 ft.²**

Wallpaper:
Conversion (width to feet): **24 in. = 2 ft.**
Formula 8-8: **A = *l* × w**
Substitution: **A = 100 ft. × 2 ft.**
Area of 1 roll: **A = 200 ft.²**

$$\left(\dfrac{420\ \cancel{ft.^2}}{}\right)\left(\dfrac{1\ roll}{200\ \cancel{ft.^2}}\right) = 2.1\ rolls$$

Solution: Buy **3** rolls.

26. 4 rolls
20-foot-long walls:
Formula 8-8: **A = *l* × w**
Substitution: **A = 7.5 ft. × 20 ft.**
Area of 1 wall: **A = 150 ft.²**
Area of 2 walls: **A =** 150 ft.² × 2 = **300 ft.²**

16-foot-long walls:
Formula 8-8: **A = *l* × w**
Substitution: **A = 8 ft. × 16 ft.**
Area of 1 wall: **A = 128 ft.²**
Area of 2 walls: **A =** 128 ft.² × 2 = **256 ft.²**

All walls:
Area of all walls: **A =** 300 ft.² + 256 ft.²
= **556 ft.²**

Wallpaper:
Conversion (width to feet): **18 in. = 1.5 ft.**
Formula 8-8: **A = *l* × w**
Substitution: **A = 100 ft. × 1.5 ft.**
Area of 1 roll: **A = 150 ft.²**

$$\left(\dfrac{556\ \cancel{ft.^2}}{}\right)\left(\dfrac{1\ roll}{150\ \cancel{ft.^2}}\right) = 3.71\ rolls$$

Solution: Buy **4** rolls.

27. 3 gal.
48-foot walls:
Formula 8-8: **A = *l* × w**

Substitution: **A = 48 ft. × 8 ft.**
Area of 1 wall: **A = 384 ft.²**
Area of 2 walls: **A** = 384 ft.² × 2 = **768 ft.²**

24-foot walls:
Formula 8-8: **A = l × w**
Substitution: **A = 24 ft. × 8 ft.**
Area of 1 wall: **A = 192 ft.²**
Area of 2 walls: **A** = 192 ft.² × 2 = **384 ft.²**

All walls:
Area of all walls: **A** = 768 ft.² + 384 ft.²
= **1152 ft.²**

$$\left(\frac{1152 \cancel{ft.^2}}{}\right)\left(\frac{1 \text{ gal.}}{400 \cancel{ft.^2}}\right) = 2.88 \text{ gal.}$$

Solution: Buy **3** gallons.

28. 15,625 ft.²
Formula 8-9: **A = s²**
Substitution: **A = 125 ft. × 125 ft.**
Solution: **A = 15,625 ft.²**

29. 5.83 m
Formula 8-9: **A = s²**
Substitution: **34 m² = s²**
Transposition: **s = $\sqrt{34 \text{ m}^2}$**
Solution: **s = 5.83 m**

30. 150 ft.²
Formula 8-10: **A = 1/2(b × h)**
Substitution: **A = 1/2 × 20 ft. × 15 ft.**
Solution: **A = 150 ft.²**

31. 144 ft.²
Formula 8-10: **A = 1/2(b × h)**
Substitution: **A = 1/2 × 24 ft. × 12 ft.**
Solution: **A = 144 ft.²**

32. 390 ft.²
Triangular portion:
Formula 8-10: **A = 1/2(b × h)**
Substitution: **A = 1/2 × 30 ft. × 6 ft.**
Area of triangular portion: **A × 90 ft.²**

Rectangular portion:
Formula 8-10: **A = l × w**
Substitution: **A = 30 ft. × 10 ft.**
Area of rectangular portion: **A = 300 ft.²**
Solution: **A** = 90 ft. + 300 ft.² = **390 ft.²**

33. 3216.99 in.²
Formula 8-11: **A = πr²**
Substitution: **A** = π × **(32 ft.)²**
Solution: **A = 3216.99 in.²**

34. 3421.19 cm²
Formula 8-12: **A = πd²/4**
Substitution: **A** = π × **(66 cm)²/4**
Solution: **A = 3421.19 cm²**

Volume

35. 24 ft.³
Formula 8-13: **V = l × w × h**
Substitution: **V = 3 ft. × 4 ft. × 2 ft.**
Solution: **V = 24 ft.³**

36. 31.5 ft.³
Formula 8-13: **V = l × w × h**
Substitution: **V = 3.5 ft. × 6 ft. × 1.5 ft.**
Solution: **V = 31.5 ft.³**

37. 840 ft.³
Formula 8-13: **V = l × w × h**
Substitution: **V = 10 ft. × 12 ft. × 7 ft.**
Solution: **V = 840 ft.³**

38. 64 m³
Formula 8-14: **V = s³**
Substitution: **V = (4 m)³** = 4 m × 4 m × 4 m
Solution: **V = 64 m³**

39. 1608.50 in.³
Formula 8-18: **V = πr² × h**
Substitution: **V** = π × **(2 in.)²** × **128 in.**
Solution: **V = 1608.50 in.³**

40. 28,720 gal.
Formula 8-13: **V = l × w × h**
Substitution: **V = 24 ft. × 16 ft. × 10 ft.**
Solution: **V** = 3840 ft.³ = **28,720 gal.**

41. 9.6 min.
Formula 8-15: **V = 1/2 (b × h × d)**
Substitution: **V = 1/2 × 30 ft. × 8 ft. × 60 ft.**
Volume of attic: **V = 7200 ft.³**
Solution: $\left(\frac{7200 \cancel{ft.^3}}{}\right)\left(\frac{1 \text{ min.}}{750 \cancel{ft.^3}}\right) =$ **9.6 min.**

ANSWERS TO TEST MASTERS, page 115

1. 750 ft.
Formula 8-2: p = 2l + 2w
Substitution: p = 2(150 ft.) + 2(225 ft.)
Solution: p = 750 ft.

2. 44 in.
Formula 8-2: p = 2l + 2w
Substitution: p = 2(10 in.) + 2(12 in.)
Solution: p = 44 in.

3. 400 m
Formula 8-3: p = 4s
Substitution: p = 4(100 m)
Solution: p = 400 m

4. 15 yd.
Formula 8-3: p = 4s
Substitution: 60 yd. = 4s
Solution: p = 15 yd.

5. 32 ft.
Formula 8-5: p = 3s
Substitution: p = 3(128 in.)
Solution: p = 384 in. = 32 ft.

6. 120 cm
Formula: p = 8s
Substitution: p = 8(15 cm)
Solution: p = 120 cm

7. 94.2 ft.
Formula 8-6: p = 2πr
Substitution: p = 2 × π × 15 ft.
Solution: p = 94.2 ft.

8. 7.85 m
Formula 8-7: p = πd
Substitution: p = π × 2.5 m
Solution: p = 7.85 m

9. 32 ft.²
Formula 8-8: A = l × w
Substitution: A = 8 ft. × 4 ft.
Solution: A = 32 ft.²

10. 304 ft.²
14-foot walls:
Formula 8-8: A = l × w
Substitution: A = 8 ft. × 14 ft.

Area of 1 wall: A = 112 $ft.^2$
Area of 2 walls: A = 112 $ft.^2$ × 2 = 224 $ft.^2$

10-foot wall:
Formula 8-8: A = l × w
Substitution: A = 8 ft. × 10 ft.
Area of 3rd wall: A = 80 $ft.^2$

All walls:
Area of all walls: A = 224 $ft.^2$ + 80 $ft.^2$ = 304 $ft.^2$

11. 3 rolls
Walls:
Formula 8-8: A = l × w
Substitution: A = 8 ft. × 12 ft.
Area of 1 wall: A = 96 $ft.^2$
Area of 4 walls: A = 96 $ft.^2$ × 4 = 384 $ft.^2$

Wallpaper:
Conversion (width to feet): 24 in. = 2 ft.
(length to feet): 25 yd. = 75 ft.
Formula 8-8: A = l × w
Substitution: A = 2 ft. × 75 ft.
Area of 1 roll: A = 150 $ft.^2$

$$\left(\frac{384\ \cancel{ft.^2}}{}\right)\left(\frac{1\ \text{roll}}{150\ \cancel{ft.^2}}\right) = 2.56\ \text{rolls}$$

Solution: Buy 3 rolls.

12. \$21.40
Formula 8-8: A = l × w
Substitution: A = 8 ft. × 12 ft.
Area of floor: A = 96 $ft.^2$
Conversion (to square yards): 96 $ft.^2$ = 10.7 $yd.^2$

Solution: $\left(\frac{10.7\ \cancel{yd.^2}}{}\right)\left(\frac{\$2}{\cancel{yd.^2}}\right) = \$21.40$

13. 1400 $ft.^2$
Formula 8-8: A = l × w
Substitution: A = 50 ft. × 28 ft.
Solution: A = 1400 $ft.^2$

14. 36 sheets
Roof:
Formula 8-8: A = l × w
Substitution: A = 48 ft. × 24 ft.
Area of roof: A = 1152 $ft.^2$

Plywood:
Formula 8-8: A = l × w
Substitution: A = 8 ft. × 4 ft.
Area of plywood: A = 32 $ft.^2$

Solution: $\left(\frac{1152\ \cancel{ft.^2}}{}\right)\left(\frac{1\ \text{sheet}}{32\ \cancel{ft.^2}}\right) = 36\ \text{sheets}$

15. \$93,600
Formula 8-8: A = l × w
Substitution: A = 52 ft. × 36 ft.
Area of floor: A = 1872 $ft.^2$

Solution: $\left(\frac{1872\ \cancel{ft.^2}}{}\right)\left(\frac{\$50}{\cancel{ft.^2}}\right) = \$93,600$

16. 1 gal.
Formula 8-10: A = 1/2 (b × h)
Substitution: A = 1/2 × 32 ft. × 10 ft.
Area of gable: A = 160 $ft.^2$

$$\left(\frac{160\ \cancel{ft.^2}}{}\right)\left(\frac{1\ \text{gal.}}{200\ \cancel{ft.^2}}\right) = 0.8\ \text{gal.}$$

Solution: Buy 1 gallon

17. 1810 in.
Formula 8-11: A = πr^2
Substitution: A = π × $(24\ in.)^2$
Solution: A = 1810 $in.^2$

18. 254.5 cm^2
Formula 8-12: A = $\pi d^2/4$
Substitution: A = π × $(18\ cm)^2/4$
Solution: A = 254.5 cm^2

19. 2688 cubes
Box:
Formula 8-13: V = l × w × h
Substitution: V = 16 in. × 14 in. × 12 in.
Volume of box: V = 2688 $in.^3$

Cube:
Formula 8-14: V = s^3
Substitution: V = $(1\ in.)^3$
Volume of cube: V = 1 $in.^3$

Solution: $\left(\frac{2688\ \cancel{in.^3}}{}\right)\left(\frac{1\ \text{cube}}{\cancel{in.^3}}\right) = 2688\ \text{cubes}$

20. 1120 $ft.^3$
Formula 8-13: V = l × w × d
Substitution: V = 10 ft. × 16 ft. × 7 ft.
Solution: V = 1120 $ft.^3$

21. 32 min.
Tent volume:
Formula 8-15: V = 1/2 (b × h × d)
Substitution: V = 1/2 × 8 ft. × 4 ft. × 4 ft.
Volume of tent: V = 64 $ft.^3$

Solution: $\left(\frac{64\ \cancel{ft.^3}}{}\right)\left(\frac{1\ \text{min.}}{2\ \cancel{ft.^3}}\right) = 32\ \text{min.}$

22. 85.5 $ft.^3$
Conversion (radius to feet): 4 in. = 0.33 ft.
Formula 8-18: V = πr^2 × h
Substitution: V = π × $(0.33\ ft.)^2$ × 250 ft.
Solution: V = 85.5 $ft.^3$

23. 0.59 $yd.^3$
Conversion (height to feet): 6 in. = 0.5 ft.
(width to feet): 24 in. = 2 ft.
Formula 8-13: V = l × w × d
Substitution: V = 16 ft. × 2 ft. × 0.5 ft.
Volume of concrete: V = 16 $ft.^3$ = 0.59 $yd.^3$

24. 9.375 cords
Formula 8-13: V = l × w × d
Substitution: V = 50 ft. × 4 ft. × 6 ft.
Volume of stack: V = 1200 $ft.^3$

Solution: $\left(\frac{1200\ \cancel{ft.^3}}{}\right)\left(\frac{1\ \text{cord}}{128\ \cancel{ft.^3}}\right) = 9.375\ \text{cords}$

25. 40 gal.
Conversion (depth to inches): 8 in. = 0.667 ft.
Formula 8-13: V = l × w × d
Substitution: V = 4 ft. × 2 ft. × 0.667 ft.
Volume of tub: V = 5.34 $ft.^3$ = 40 gal.

Chapter 9
READING TOOLS OF MEASUREMENT

Objectives

After studying this chapter, the student will be able to:

- Explain basic terminology of some common tools of measurement.
- Read and use a variety of tape measures.
- Read and use architects', engineers', and metric scales.
- Read and use standard-inch, metric, and vernier micrometers.
- Read a multimeter.

ANSWERS TO TEXT, page 133

U.S. Conventional Rules

1. a = 1′1 5/8″
   b = 1′2 1/4″
   c = 1′3 3/8″
2. a = 11 9/16″
   b = 1′3/8″
   c = 1′2 11/16″
3. a = 11 9/32″
   b = 11 21/32″
   c = 12 7/8″
4. a = 13.4″
   b = 14.5″
   c = 15.7″
5. a = 0.92′
   b = 1.05′
   c = 1.1′

SI Metric Rules

6. a = 2 cm
   b = 6.5 cm
   c = 10.2 cm
7. a = 5.5 cm
   b = 15.9 cm
   c = 76 mm

Triangular Drafting Scales

8. a = 7′
   b = 17′10″
   c = 1′1″
9. a = 2′4 3/4″
   b = 3′4″
   c = 7′9″
10. a = 2.5′
    b = 5.2′
    c = 2.8′
11. a = 2′6 1/2″
    b = 3′5″
    c = 1 7/16″

Micrometers

12. Sleeve = 0.600″
    Thimble = 0.011″
    Measurement = 0.611″
13. Sleeve = 0.125″
    Thimble = 0.016″
    Measurement = 0.141″
14. Sleeve = 0.900″
    Thimble = 0.002″
    Measurement = 0.902″
15. Sleeve = 0.000″
    Thimble = 0.003″
    Measurement = 0.003″
16. Sleeve = 0.500″
    Thimble = 0.000″
    Measurement = 0.500″
17. Sleeve = 14.50 mm
    Thimble = 0.31 mm
    Measurement = 14.81 mm
18. Sleeve = 5.00 mm
    Thimble = 0.22 mm
    Measurement = 5.22 mm
19. Sleeve = 23.50 mm
    Thimble = 0.18 mm
    Measurement = 23.68 mm
20. Sleeve = 0.30 mm
    Thimble = 0.04 mm
    Measurement = 0.34 mm
21. Sleeve = 16.50 mm
    Thimble = 0.48 mm
    Measurement = 16.98 mm
22. Sleeve = 0.825″
    Thimble = 0.007″
    Vernier = 0.0006″
    Measurement = 0.8326″
23. Sleeve = 0.350″
    Thimble = 0.005″
    Vernier = 0.0002″
    Measurement = 0.3552″
24. Sleeve = 0.050″
    Thimble = 0.012″
    Vernier = 0.0001″
    Measurement = 0.0621″
25. Sleeve = 0.225″
    Thimble = 0.017″
    Vernier = 0.0006″
    Measurement = 0.2426″

Electrical Multimeters: Volt-Ohm-Milliammeters

26. range = 10 ACV
    scale = 0, 2, 4, 6, 8, 10 ACV
    line = 0.2 ACV
27. a = 2.4 ACV
    b = 5.4 ACV
    c = 8.8 ACV
28.

NEEDLE	SMALL LINE	READING	RANGE	MEASUREMENT
a	20 Ω	140 Ω	x 1	140 Ω
b	2 Ω	32 Ω	x 1	32 Ω
c	0.5 Ω	8.5 Ω	x 1	8.5 Ω

29. range = 2.5 DCmA
    scale = 0, 0.5, 1.0, 1.5, 2.0, 2.5 DCmA
    line = 0.05 DCmA
30. a = 0.4 mA
    b = 1.05 mA
    c = 1.95 mA
31. range = 250 DCV
    scale = 0, 50, 100, 150, 200, 250 DCV
    line = 5 DCV
32. a = 75 DCV
    b = 115 DCV
    c = 180 DCV

ANSWERS TO WORKBOOK, PAGE 81

U.S. Conventional Rules

1. a = **1′5/8″**
   b = **1′1 3/8″**
   c = **1′ 2 1/4″**
2. a = **13 1/2″**
   b = **14 9/16″**
   c = **15 3/8″**
3. a = **1′1 3/16″**
   b = **11 11/32″**
   c = **13 7/8″**
4. a = **0.85′**
   b = **0.99′**
   c = **1.1′**
5. a = **1.07′**
   b = **11 19/32″**
   c = **1′1 5/16″**

SI Metric Rules

6. a = **2.7 cm**
   b = **5.1 cm**
   c = **10 cm**
7. a = **3.2 cm**
   b = **10.7 cm**
   c = **79 mm**

Triangular Drafting Scales

8. a = **2′7 1/2″**
   b = **2′**
   c = **14′6″**
9. a = **7′**
   b = **23′**
   c = **390′**
10. a = **1′4″**
    b = **12′10 1/2″**
    c = **5′7″**
11. a = **19′**
    b = **7′3″**
    c = **3′7 1/2″**

Micrometers

12. Sleeve = **0.300″**
    Thimble = **0.006″**
    Measurement = **0.306″**
13. Sleeve = **0.125″**
    Thimble = **0.012″**
    Measurement = **0.137″**
14. Sleeve = **0.850″**
    Thimble = **0.024″**
    Measurement = **0.874″**
15. Sleeve = **0.225″**
    Thimble = **0.023″**
    Measurement = **0.248″**
16. Sleeve = **0.475″**
    Thimble = **0.021″**
    Measurement = **0.496″**
17. Sleeve = **0.400″**
    Thimble = **0.008″**
    Measurement = **0.408″**
18. Sleeve = **0.225″**
    Thimble = **0.014″**
    Measurement = **0.239″**
19. Sleeve = **0.075″**
    Thimble = **0.004″**
    Measurement = **0.079″**
20. Sleeve = **0.450″**
    Thimble = **0.009″**
    Measurement = **0.459″**
21. Sleeve = **0.975″**
    Thimble = **0.007″**
    Measurement = **0.982″**
22. Sleeve = **20.50 mm**
    Thimble = **0.47 mm**
    Measurement = **20.97 mm**
23. Sleeve = **8.00 mm**
    Thimble = **0.08 mm**
    Measurement = **8.08 mm**
24. Sleeve = **2.00 mm**
    Thimble = **0.20 mm**
    Measurement = **2.20 mm**
25. Sleeve = **16.00 mm**
    Thimble = **0.37 mm**
    Measurement = **16.37 mm**
26. Sleeve = **11.50 mm**
    Thimble = **0.19 mm**
    Measurement = **11.69 mm**
27. Sleeve = **0.150″**
    Thimble = **0.015″**
    Vernier = **0.0003″**
    Measurement = **0.1653′**
28. Sleeve = **0.450″**
    Thimble = **0.023″**
    Vernier = **0.0003″**
    Measurement = **0.4733″**
29. Sleeve = **0.350″**
    Thimble = **0.002″**
    Vernier = **0.0000″**
    Measurement = **0.3520″**
30. Sleeve = **0.850″**
    Thimble = **0.021″**
    Vernier = **0.0007″**
    Measurement = **0.8717″**

Electrical Multimeters: Volt-Ohm-Milliammeters

31. range = **1000 ACV**
    scale = 0, **200, 400, 600, 800, 1000 ACV**
    line = **20 ACV**
32. a = **240 ACV**
    b = **480 ACV**
    c = **940 ACV**
33. range = **50 DCV**
    scale = 0, **10, 20, 30, 40, 50 DCV**
    line = **1 DCV**
34. a = **8 DCV**
    b = **20.5 DCV**
    c = **39 DCV**
35.

NEEDLE	SMALL LINE	READING	RANGE	MEASUREMENT
a	5 Ω	70 Ω	× 1k	70,000 Ω
b	1 Ω	17 Ω	× 1k	17,000 Ω
c	0.5 Ω	3.5 Ω	× 1k	3500 Ω

36.
    range = **0.5 DCV**
    scale = 0, **0.1, 0.2, 0.3, 0.4, 0.5 DCV**
    line = **0.01 DCV**
37. a = **0.05 DCV**
    b = **0.23 DCV**
    c = **0.39 DCV**
38. range = **25 DCmA**
    scale = 0, **5, 10, 15, 20, 25 DCmA**
    line = **0.5 DCmA**
39. a = **5.3 DCmA**
    b = **11.7 DCmA**
    c = **19 DCmA**

ANSWERS TO TEST MASTERS, page 122

1. a = **2′ 11 3/8″**
   b = **3′ 1/4″**
   c = **3′ 1 3/4″**
2. a = **11 3/16″**
   b = **12 5/8″**
   c = **13 5/16″**
3. a = **11 7/32″**
   b = **12 7/16″**
   c = **13 7/8″**
4. a = **7.3″**
   b = **8.0″**
   c = **8.8″**
5. a = **1.23′**
   b = **1.35′**
   c = **1.46′**
6. a = **2.7 cm**
   b = **5.0 cm**
   c = **8.3 cm**
7. a = **6.9 cm**
   b = **181 mm**
   c = **55 mm**
8. a = **1 11/16″**
   b = **5′ 10″**
   c = **2′ 5″**
9. a = **1′ 4 1/2″**
   b = **4′ 4″**
   c = **17′ 8″**
10. a = **2.3′**
    b = **17′**
    c = **45′**
11. a = **11′ 2″**
    b = **5′ 3″**
    c = **3′ 7″**
12. Sleeve = **0.075′**
    Thimble = **0.000″**
    Measurement = **0.075″**
13. Sleeve = **0.100″**
    Thimble = **0.022″**
    Measurement = **0.122″**
14. Sleeve = **0.250″**
    Thimble = **0.019″**
    Measurement = **0.269″**
15. Sleeve = **0.325″**
    Thimble = **0.013″**
    Measurement = **0.338″**
16. Sleeve = **0.850″**
    Thimble = **0.005″**
    Measurement = **0.855″**
17. Sleeve = **24.00 mm**
    Thimble = **0.30 mm**
    Measurement = **24.30 mm**
18. Sleeve = **12.50 mm**
    Thimble = **0.21 mm**
    Measurement = **12.76 mm**
19. Sleeve = **18.50 mm**
    Thimble = **0.00 mm**
    Measurement = **18.50 mm**
20. Sleeve = **5.00 mm**
    Thimble = **0.14 mm**
    Measurement = **5.14 mm**
21. Sleeve = **22.50 mm**
    Thimble = **0.10 mm**
    Measurement = **22.60 mm**
22. Sleeve = **0.525″**
    Thimble = **0.000″**
    Vernier = **0.0003″**
    Measurement = **0.5253″**
23. Sleeve = **0.925″**
    Thimble = **0.011″**
    Vernier = **0.0000″**
    Measurement = **0.9360″**
24. Sleeve = **0.175″**
    Thimble = **0.014″**
    Vernier = **0.0006″**
    Measurement = **0.1896″**
25. Sleeve = **0.675″**
    Thimble = **0.022″**
    Vernier = **0.0001″**
    Measurement = **0.6971″**
26. range = **50 ACV**
    scale = **0, 10, 20, 30, 40, 50 ACV**
    line = **1 ACV**
27. a = **9 ACV**
    b = **27 ACV**
    c = **45 ACV**

28.

NEEDLE	SMALL LINE	READING	RANGE	MEASUREMENT
a	2 Ω	44 Ω	× 10k	440,000 Ω
b	2 Ω	22 Ω	× 10k	220,000 Ω
c	0.5 Ω	7 Ω	× 10k	70,000 Ω

29. range = **250 DCmA**
scale = **0, 50, 100, 150, 200, 250 DCmA**
line = **5 DCmA**
30. a = **75 DCmA**
b = **130 DCmA**
c = **212 DCmA**
31. range = **2.5 DCV**
scale = **0, 0.5, 1.0, 1.5, 2.0, 2.5 DCV**
line = **0.05 DCV**
33. a = **0.19 DCV**
b = **1.25 DCV**
c = **2.04 DCV**

Chapter 10
SIGNED NUMBERS AND OPERATIONS

Objectives
After studying this chapter, the student will be able to:
- Explain basic terminology concerning signed numbers, powers, and roots.
- Determine if "signed" fractions are positive or negative.
- Perform operations of addition, subtraction, multiplication, and division using signed numbers.
- Find the solutions of numbers raised to powers.
- Find square and cube roots of numbers.
- Evaluate expressions with more than one operation by following the correct order of operations.

ANSWERS TO TEXT, page 154

The Number Line
1. +50
2. −68
3. +20
4. −10
5. +20
6. 8
7. 14
8. 1
9. 10
10. 5
11. <
12. >
13. <
14. <
15. >

Positive and Negative Fractions
16. −
17. +
18. +
19. −
20. +
21. −
22. −
23. +
24. −
25. +

Addition and Subtraction of Signed Numbers
26. −3
27. 12
28. 4
29. −6
30. −2
31. 18
32. −5
33. −14
34. $1\frac{1}{6}$
35. $-\frac{1}{4}$
36. $\frac{1}{8}$
37. $\frac{1}{2}$
38. $-\frac{3}{4}$
39. $-1\frac{5}{12}$
40. −29.55
41. 4.01
42. −4
43. −20
44. −11
45. 17

Multiplication and Division of Signed Numbers
46. 15
47. −24
48. −3.15
49. 65.1
50. −5.628
51. −15.87
52. 0
53. 50
54. $-\frac{1}{2}$
55. $\frac{5}{18}$
56. 2
57. −6
58. 5
59. 2
60. −3

61. 3
62. −3
63. 7.5
64. −6
65. −7.5

Word Problems with Signed Numbers

66. 1 mi.

Assign: east = +
west = −

+5 miles (east)
−8 miles (west)
+2 miles (east)

$+5 + (-8) + 2 = -1$ mile = 1 mile west

67. $370.14

+
250.50
125.00
225.60
$601.10 (total positive)

−
23.09
109.02
1.45
75.65
21.75
$230.96 (total negative)

601.10 − 230.96 = $370.14

68. $-2.14°$

$-15 + (-10) + 3 + 8 + (-2) + 1 + 0 = -15°$

$$\frac{-15}{7} = -2.14°$$

69. 48 in.

$$\left(\frac{8\ \cancel{\text{ft.}}}{}\right)\left(\frac{12\ \text{in.}}{1\ \cancel{\text{ft.}}}\right) = 96\ \text{in. board}$$

original board = 96 in.
pieces cut = (15 × 5) = 75 in.
piece left over = 96 − 75 = 21 in.

21 + 27 = 48 in. length for next application

70. 14 books/box

Books remaining = 9
Books short = 5 × 3 = 15
Books removed = 6 × 3 = 18

9 + 15 + 18 = 42 books
42 ÷ 3 = 14 books/box

71. $-12°$
$+18 - 30 = -12°$
72. −$250 net income (or $250 net loss)
+1500 − 1750 = −$250 net income
73. 29 feet deep
−52 + 23 = −29 feet from surface
(or 29 feet deep)
74. $60 net loss (or −$60 net profit)
−320 + (−320 × 2) = −$960 (or $960 paid)
−960 + 900 = −$60 net profit
75. $27,000 gross
−9000 + (−6000) = −$15,000
−6000 × 2 = −$12,000
−15,000 + (−12,000) = −$27,000
−27,000 + 27,000 = $0

Powers and Roots

76. 36
77. 64
78. 27
79. 5
80. 7
81. −2
82. 4.5
83. 6.3
84. 0.45

Order of Operations

85. 2
86. 3
87. 4
88. −44
89. −23
90. −4
91. −35
92. 6
93. 16
94. 17
95. −3

ANSWERS TO WORKBOOK, page 95

The Number Line

1. −35
2. −56
3. +32
4. +15
5. −15
6. 23
7. 8
8. 0
9. 3
10. 6
11. >
12. >
13. <
14. <
15. >

Positive and Negative Fractions

16. −
17. +
18. +
19. −
20. −
21. −
22. +
23. +
24. −
25. +

Addition and Subtraction of Signed Numbers

26. −6
27. 5

28. -18
29. -17
30. 8
31. -13
32. -5
33. 15
34. $-1\frac{5}{12}$
35. $-\frac{1}{8}$
36. $1\frac{5}{9}$
37. $\frac{5}{6}$
38. -23.58
39. 14.11
40. 23.3
41. 0.312
42. 0
43. -25
44. 17
45. 35,406

Multiplication and Division of Signed Numbers

46. -60
47. 49
48. -0.48
49. 0.025
50. 0.008
51. 90
52. 168
53. $-\frac{8}{35}$
54. $1\frac{23}{32}$
55. $-\frac{7}{12}$
56. -8
57. -3
58. 2
59. -6
60. 5
61. 10
62. -3
63. 5
64. -2
65. 29
66. \$2300 unpaid balance (or $-$\$2300 balance)

car purchase = $-$\$10,000
down payment = +\$2000
monthly payments = (+\$350 × 2) = +\$700
payment = +\$5000

$-10,000 + 2000 + 700 + 5000 = -\$2300$

67. \$441.86

+
350.05
125.60
251.20 (2 deposits @ \$125.60 each)
\$726.85 (total positive)

−
69.25
21.46
90.50
35.80
1.05
26.93
40.00 (4 checks @ \$10.00 each)
\$284.99 (total negative)

726.85 − 284.99 = \$441.86

68. $-5.71°$
$-15 + (-25) + (-10) + 0 + 8 + 7 + (-5) = -40°$

$\frac{-40}{7} = -5.71°$

69. 60 2 × 4s
250 − 60 − 80 + 300 − 150 − 100 − 100 = 60 2 x 4s
70. 5 passes
1.00 − 0.75 = 0.25″ removed

$\left(\frac{0.25'' \text{ ~~removed~~}}{}\right)\left(\frac{1 \text{ pass}}{0.05'' \text{ ~~removed~~}}\right) = 5 \text{ passes}$

Powers and Roots

71. 4
72. 16
73. -125
74. 0.09
75. -0.001
76. 1
77. 2
78. 10
79. 4
80. -1
81. 3.2
82. 4.2
83. 7.1
84. 0.77

Order of Operations

85. -8
86. -17
87. -23
88. -49
89. -3
90. 3
91. -6
92. 3
93. -26
94. 33
95. -10

ANSWERS TO TEST MASTERS, page 131

1. −25
2. +560
3. +98.6
4. −3
5. −5
6. 10
7. 2.2
8. >
9. >
10. >
11. +
12. +
13. −
14. −
15. −
16. +
17. −2
18. 6
19. −12
20. −10
21. 1
22. 5
23. −9
24. 12
25. 4
26. $-\frac{1}{2}$
27. −30
28. −6
29. $-\frac{2}{5}$
30. $\frac{5}{63}$
31. 3
32. −4
33. −7
34. −10
35. −0.2
36. −0.2
37. 6
38. 29
39. 18
40. 0
41. 49
42. −64
43. 7
44. −3
45. 7.7
46. 9.5
47. 41
48. −4
49. 2
50. 25 lb.
   50 −3(5) + 2 + 3 − 1 − 3(7) + 1 + 4 + 2 = 25 lb.

Chapter 11
INTRODUCTION TO ALGEBRA

Objectives

After studying this chapter, the student will be able to:
- Explain basic terminology of algebra.
- Perform operations of addition, subtraction, multiplication, and division on monomials and polynomials.
- Simplify algebraic expressions by the laws of exponents.
- Simplify algebraic expressions by factoring.

ANSWERS TO TEXT, page 165

Basic Terminology

term	coefficient	variables	constant	constant term
1. $-5x$ $+6y^2$ -7	−5 +6	x y	−5 +6 −7	−7
2. $-3x$ $-4xy^2$ $+15$	−3 −4	x x,y	−3 −4 +15	+15
3. $+9$ $+8u$ $-v$	 +8 −1	 u v	+9 +8 −1	+9
4. $+x^3$ $-3x^2$ $+8$ $-12y$ $+z$	+1 −3 −12 +1	x x y z	+1 −3 +8 −12 +1	+8
5. $-3x^2y^3$ -10	−3	x,y	−3 −10	−10

term	factors
6. $-8a^3$ $+3ab$ -4	$-8\cdot a\cdot a\cdot a$ $+3\cdot a\cdot b$
7. $+5$ $-(3)(4x)$ $+2x^2$ -10	 $-1\cdot 3\cdot 4\cdot x$ $+2\cdot x\cdot x$
8 $+6(x+1)$ $+4$ $-3xy$	$+6\cdot(x+1)$ $-3\cdot x\cdot y$
9. $+10$ $-2(x+y)$ $+(3x-2)$	 $-2\cdot(x+y)$ $+1\cdot(3x-2)$
10. $x^2y^3z^2$ $+3(a+2b)$	$+1\cdot x\cdot x\cdot y\cdot y\cdot y\cdot z\cdot z$ $+3\cdot(a+2b)$

Operations

11. $-2a + 3y$
12. $3x^2 - 7x + 2$

13. 0
14. b
15. $-d - 7$
16. 0
17. $-7 - 4c$
18. $-x^2y - xy$
19. $6xy$
20. $-3a + 7abc$
21. $-15a^2 + 1$
22. $12x^9$
23. $2xyz$
24. $-\frac{1}{2y^2}$
25. $\frac{2}{y^2}$
26. -1
27. b
28. x^8y^{12}
29. $x^2y^2z^2$
30. $\frac{1}{y^4}$
31. $6a + b$
32. $-7x - 3y$
33. $a + 4c$
34. $4y + 2z - 5$
35. $4x + 4y - 3z$
36. $4x + 4y - 14xy + 1$
37. $8x + y - 1$
38. $2x^3 - 1$
39. $-c - 3d$
40. $-5x - 2y$
41. $8a^2b^4c$
42. $-6x^3y^5$
43. $24b^3c$
44. $2a + 8b - 2c - 6d$
45. $-8a^3 + 4a^2 + 12a^2b$
46. $2a^2b - 6ab^2$
47. $-a + 5b$
48. $18a + 30b$
49. $3x^2 + 5xy - 2y^2$
50. $-4xy - 12x^2$
51. $6x + 4$
52. $4b - 1$
53. $x^3 - x$
54. $3xy - 2x + y$
55. $x - 2$
56. $3x(1 + y)$
57. $3(x + 2y)$
58. $(8x^2 + 7y^2)(8x^2 - 7y^2)$
59. $(2x + 1)^2$
60. $(x - 5)(x - 3)$
61. $\frac{1}{7x}$
62. 2
63. $1 - \frac{5}{x}$
64. $(x - 3)(x^2 + 9)$
65. $\frac{x - 2}{x + 2}$

ANSWERS TO WORKBOOK, page 107

term	coefficient	variables	constant	constant term
1. $+3x$	$+3$	x	$+3$	
-8			-8	-8
$+x^2$	$+1$	x	$+1$	
2. $+10$			$+10$	$+10$
$-5x^3$	-5	x	-5	
$+2y$	$+2$	y	$+2$	
3. $-3x^4$	-3	x	-3	
-213			-213	-213
$+3xy$	$+3$	x,y	$+3$	
4. $+4t$	$+4$	t	$+4$	
$-3r^2s$	-3	r,s	-3	
$+3u$	$+3$	u	$+3$	
5. -29			-29	-29
$+3uvw$	$+3$	u,v,w	$+3$	
$-6w^2$	-6	w	-6	

term	factors
6. $+10$	
$-5a^2b$	$-5\cdot a\cdot a\cdot b$
$+3(2 - a)$	$+3\cdot(2 - a)$
7. $+2x$	$+2\cdot x$
$-(3)(-4)$	$-1\cdot 3\cdot -4$
$+xyz$	$+1\cdot x\cdot y\cdot z$
8. $+x^3$	$+1\cdot x\cdot x\cdot x$
$-3y(x + y)$	$-3\cdot y\cdot(x + y)$
$+5$	
9. $+3x^2yz^2$	$+3\cdot x\cdot x\cdot y\cdot z\cdot z$
$+8$	
-5^3	$-1\cdot 5\cdot 5\cdot 5$
10. $+7^4$	$+7\cdot 7\cdot 7\cdot 7$
$-(-6)(-3x)$	$-1\cdot -6\cdot -3\cdot x$
$+5(x + 3)$	$+5\cdot(x + 3)$

Operations

11. $-8x + 5$
12. $3y^3 + y^2$
13. $-4x^2y + 2xy^2$
14. $-2ab - 2a^2 + 1$
15. $-4a$
16. $-3xy$
17. $-x + 30$
18. $10a^2 - ab$
19. $6x^2 + 20x + 4xy$
20. 1
21. $12x^2y$
22. $16x^{10}$
23. $48b^7$
24. x^4y^6
25. $a^3b^6c^3$
26. $-\frac{1}{3}$

27. $12y^3z$
28. $-2x^2$
29. $\frac{a}{2b}$
30. $\frac{y}{x^4}$
31. $x + 5$
32. $7a - b$
33. $-2a + 4b$
34. $-9x - 14$
35. $-10 + 15a + 15b - 45c$
36. $-21x + 7xy + 5y + 8y^2$
37. $4x^2 + 3x$
38. $3a^6b^6c^4$
39. $10x^7y^5z^4$
40. $-5a^2b + 8ab - 4$
41. $x + 2y$
42. $x^3 - 2x^2y + xy^2$
43. $-4x^4y + y$
44. $x + 4$
45. $x - 5$

Factoring

46. $2x(5 - y)$
47. $3axy^2(2a - 3x + 4a^2xy)$
48. $(2b - 3c)(2b + 3c)$
49. $(x + 9)(x - 4)$
50. $(x + 4y)^2$
51. $\frac{x^2}{x + 2}$
52. $-(x + 1)$
53. $\frac{x - 5}{x + 5}$
54. $(x^2 + 4)(x - 2)$
55. $-\frac{1}{2}$

ANSWERS TO TEST MASTERS, page 140

term	coefficient	variables	constant	constant term
1. $+4x^2$ -3 $+xy$	$+4$ $+1$	x x,y	$+4$ -3 $+1$	-3
2. $+6$ $-5uv$ $+u^2$	 -5 $+1$	 u,v u	$+6$ -5 $+1$	$+6$
3. -10 $-3x^2y^2$ $+2xy$	 -3 $+2$	 x,y x,y	-10 -3 $+2$	-10
4. $(4)(3x)$ $-y^2$ $+1$	$+12$ -1	x y	$+12$ -1 $+1$	$+1$
5. $+6x^2$ $-3xyz$ -13	$+6$ -3	x x,y,z	$+6$ -3 -13	-13

term	factors
6. $+7$ $+4a^2b$ $-3ab^2$	 $+4 \cdot a \cdot a \cdot b$ $-3 \cdot a \cdot b \cdot b$
7. $-5(x + 1)$ $-3y$ $+6xyz$	$-5 \cdot (x + 1)$ $-3 \cdot y$ $+6 \cdot x \cdot y \cdot z$
8. $(-3)(-2x)$ $-4x^2$ $+2^3$	$-3 \cdot -2 \cdot x$ $-4 \cdot x \cdot x$ $+2 \cdot 2 \cdot 2$
9. $+5$ $+(3x)(xy^2)$ $-10y$	 $+1 \cdot 3 \cdot x \cdot x \cdot y \cdot y$ $-10 \cdot y$
10. $+4^2$ $+x^3y$ $-(6x)(7x)$	$+4 \cdot 4$ $+1 \cdot x \cdot x \cdot x \cdot y$ $-1 \cdot 6 \cdot x \cdot 7 \cdot x$

11. $-10x + 13$
12. $13a^2 - ac$
13. $16x^6$
14. a^6b^8
15. $\frac{x}{2y}$
16. $4x + 13$
17. $-10x - 32$
18. $3y^2 + 5xy - 16x - 7y$
19. $4a^9b^9c^9$
20. $-4a^3b - 3a^2b + 20ab - 10$
21. $x + 7$
22. $2(x - 6)(x - 9)$
23. $(5x - 6y)(5x + 6y)$
24. $\frac{x^2}{x + 9}$
25. $\frac{(x^2 + 9)(x - 3)}{(x - 2)}$

Chapter 12
ALGEBRAIC EQUATIONS AND INEQUALITIES

Objectives

After studying this chapter, the student will be able to:

- Solve algebraic equations.
- Solve algebraic inequalities.
- Solve simultaneous equations by the addition/subtraction method.
- Solve simultaneous equations by the substitution method.

ANSWERS TO TEXT, page 185

Isolating the Unknown

1. $x = 5$
2. $a = 4$
3. $y = -10$
4. $b = 5$

5. $z = 12$
6. $c = 5$
7. $a = 2$
8. $x = 1$
9. $d = \frac{1}{2}$
10. $y = 80$
11. $x = -10$
12. $a = 1$
13. $x = -4$
14. $x \times -8$
15. $y = -2\frac{1}{2}$
16. $a = -2$
17. $y = -1$
18. $x = 1$
19. $a = 7$
20. $x = 3$
21. $x = 10$
22. $x = 2$
23. $x = 5$
24. $a = 2$
25. $y = 100$

Inequalities

26. $x > 5$
27. $y < 4$
28. $a \geq 5$
29. $b \leq 4$
30. $a < 1$

Equations with Two Unknowns

31. $x = 2, y = 4$
32. $x = 4, y = 2$
33. $x = 2, y = 1$
34. $a = 5, b = 2$
35. $x = 1, y = 1$
36. $x = 3, y = 2$
37. $x = 1, y = 5$
38. $x = -2, y = 6$
39. $x = 4, y = -1$
40. $x = -2, y = -8$

ANSWERS TO WORKBOOK, page 117

Isolating the Unknown

1. $y = -36$
2. $a = 2$
3. $b = -1$
4. $y = 1$
5. $b = 8\frac{1}{2}$
6. $z = -3$
7. $x = 3$
8. $y = 3$
9. $d = -1$
10. $x = 16$
11. $y = \frac{1}{17}$
12. $x = 2$
13. $y = -7$
14. $a = \frac{2}{3}$
15. $x = 1$
16. $x = 2$
17. $x = -3$
18. $x = -2$
19. $a = 0.1$
20. $y = 10$
21. $x = 7$
22. $y = 10$
23. $a = 81$
24. $x = 6$
25. $x = 8$

Inequalities

26. $y < 15$
27. $a > -13$
28. $b \geq 0$
29. $x \geq 1$
30. $b < 1$

Equations with Two Unknowns/Systems of Equations

31. $x = \mathbf{4}, y = \mathbf{3}$
32. $x = \mathbf{3}, y = \mathbf{2}$
33. $u = \mathbf{4}, v = \mathbf{3}$
34. $x = \mathbf{3}, y = \mathbf{3}$
35. $x = \mathbf{4}, y = \mathbf{3}$
36. $x = \mathbf{6}, y = \mathbf{4}$
37. $x = \mathbf{3}, y = \mathbf{5}$
38. $x = \mathbf{1}, y = \mathbf{7}$
39. $x = \mathbf{2}, y = \mathbf{2}$
40. $x = \mathbf{0}, y = \mathbf{-2}$

ANSWERS TO TEST MASTERS, page 146

1. $a = -13$
2. $x = -3$
3. $b = 5$
4. $x = 4$
5. $y = 6$
6. $d = -1$
7. $x = -32$
8. $x = 1$
9. $a = -9$
10. $x = 2$
11. $x = -1$
12. $x = -9$
13. $a = -1$
14. $x = 3$
15. $a = 8$
16. $x = 7$
17. $x = 9$
18. $x = 28$
19. $a < 18$
20. $b \geq 3$

21. 7, 2

$$\begin{array}{l} 5b - c = 33 \\ \underline{7b + c = 51} \\ 12b \quad = 84 \end{array}$$

$$\frac{12b}{12} = \frac{84}{12} \qquad b = 7$$

$7(7) + c = 51$
$49 + c - 49 = 51 - 49 \qquad c = 2$

22. 5, 3

$$\begin{array}{l} 2x + y = \quad 13 \\ \underline{5x + y = \quad 28} \\ -3x \quad = -15 \end{array}$$

$$\frac{-3x}{-3} = \frac{-15}{-3} \qquad x = 5$$

$2(5) + y = 13$
$10 + y - 10 = 13 - 10 \qquad y = 3$

23. 2, −4
$2x + y - 2x = 0 - 2x$
$y = -2x$
$3x - (-2x) = 10$

$$\frac{5x}{5} = \frac{10}{5} \qquad x = 2$$

$2(2) + y = 0$
$4 + y - 4 = 0 - 4 \qquad y = -4$

24. 5, 3
$x + 5y - 5y = 20 - 5y$
$x = 20 - 5y$
$y - (20 - 5y) = -2$
$6y - 20 + 20 = -2 + 20$

$$\frac{6y}{6} = \frac{18}{6} \qquad y = 3$$

$x + 5(3) = 20$
$x + 15 - 15 = 20 - 15 \qquad x = 5$

Chapter 13
GRAPHS OF LINEAR EQUATIONS AND INEQUALITIES

Objectives

After studying this chapter, the student will be able to:

- Explain basic terminology of the rectangular coordinate system.
- Draw graphs of one-variable equations.
- Draw graphs of two-variable equations.
- Give the equation of a straight line from its graph.
- Solve simultaneous equations graphically.
- Solve inequalities graphically.

ANSWERS TO TEXT, PAGE 202

1. a. (2, 1) b. (3, 5)
c. (−3, 4) d. (−5, 2)
e. (−4, −2) f. (−6, −4)
g. (−4, −6) h. (2, −6)
i. (3, −4) j. (6, −3)
k. (0, −2) l. (5, 0)

2. (0, −7) $(3\frac{1}{2}, 0)$ (2, −3): optional values

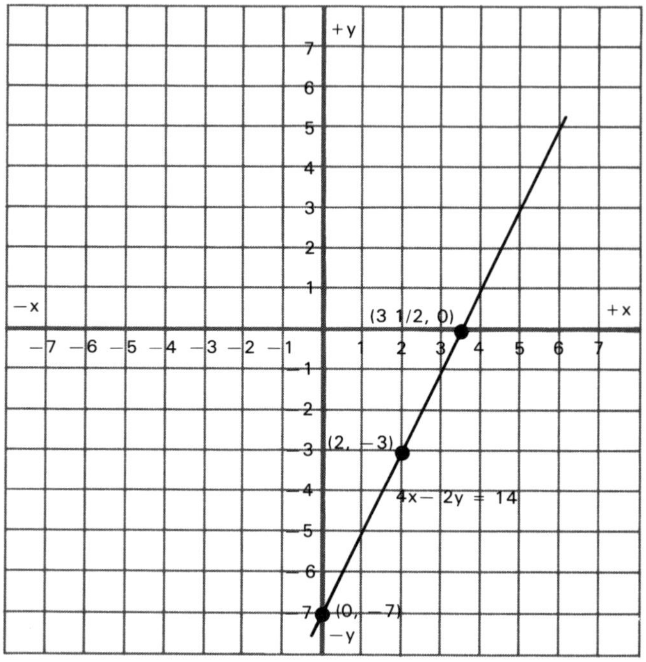

slope-intercept form of equation: $y = 2x - 7$
slope = 2 y-intercept = −7

3. equation: $y = x + 3$
(0, 3) (−3, 0): optional values

equation: $x + y = 6$
(0, 6) (6, 0): optional values

solution: $(1\frac{1}{2}, 4\frac{1}{2})$

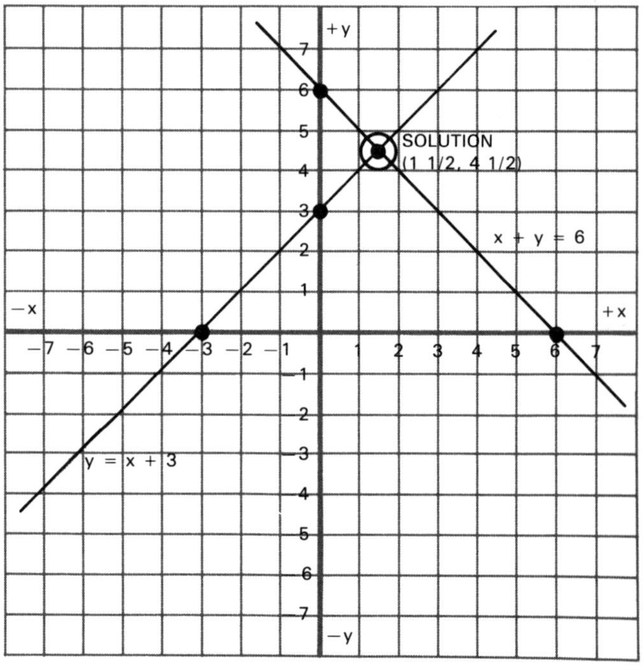

4. equation: $x = y + 3$
   $(3, 0)$ $(0, -3)$: optional values
   equation: $2x + y = 6$
   $(0, 6)$ $(3, 0)$: optional values
   solution: $(3, 0)$

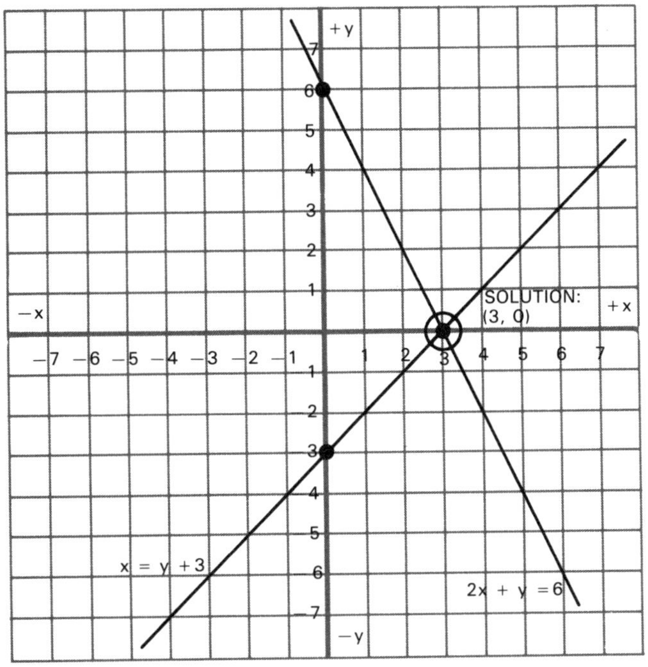

5. $(0, 5)$ $(2\frac{1}{2}, 0)$: optional values

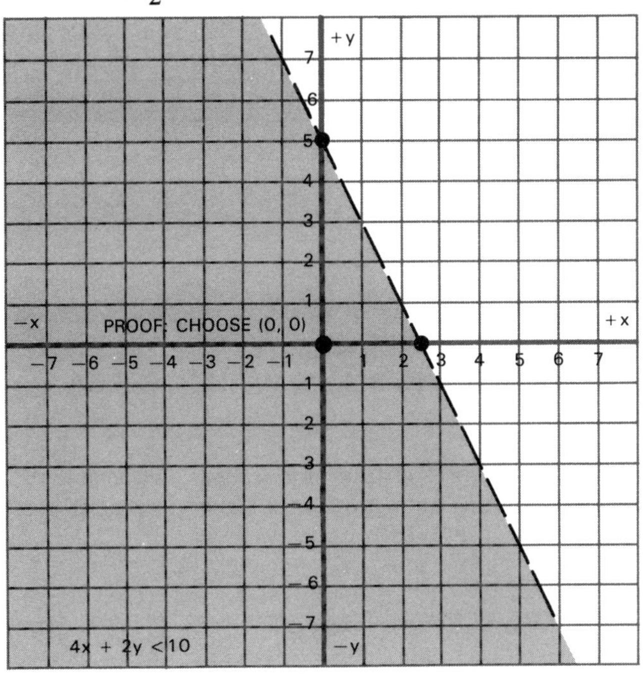

ANSWERS TO WORKBOOK, page 127

1. a. $(4, 0)$
   b. $(6, 3)$
   c. $(2, 5)$
   d. $(0, 3)$
   e. $(-1, 4)$
   f. $(-5, 4)$
   g. $(-4, -3)$
   h. $(-3, -6)$
   i. $(2, -5)$
   j. $(6, -3)$
2. $(0, 1\frac{1}{3})(-2, 0)(1, 2)$: optional values

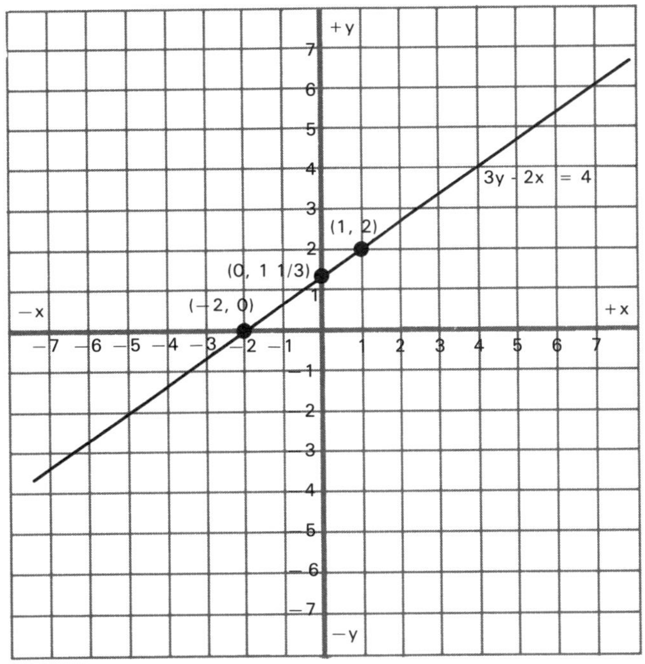

slope-intercept form of equation: $y = \frac{2}{3}x + \frac{4}{3}$

slope $= \frac{2}{3}$ y-intercept $= \frac{4}{3}$

3. equation: $3x - 5y = 2$
   $(0, -\frac{2}{5})$ $(\frac{2}{3}, 0)$ $(4, 2)$: optional values
   equation: $4x - 6 = 5y$
   $(0, -1\frac{1}{5})$ $(1\frac{1}{2}, 0)$ $(4, 2)$: optional values
   solution: $(4, 2)$

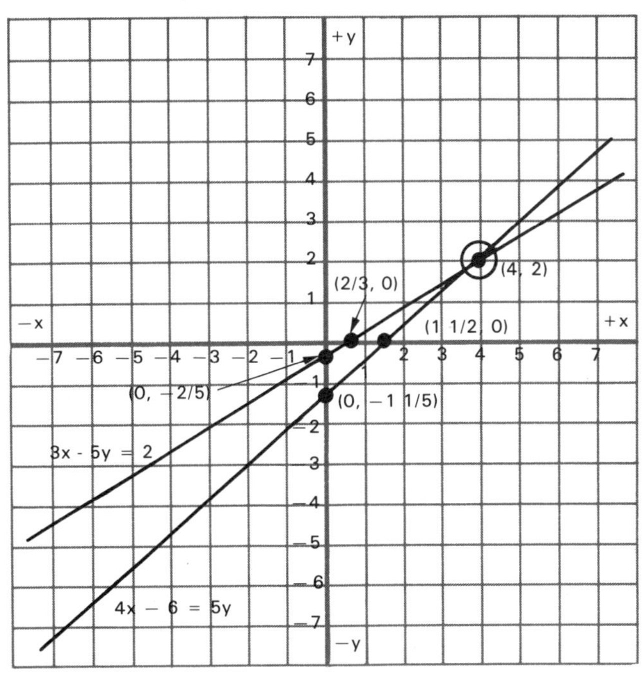

4. equation: x + y = 6
   (0, 6) (6, 0) (4, 2): optional values
   equation: x − y = 2
   (0, −2) (2, 0) (4, 2): optional values
   solution: (4, 2)

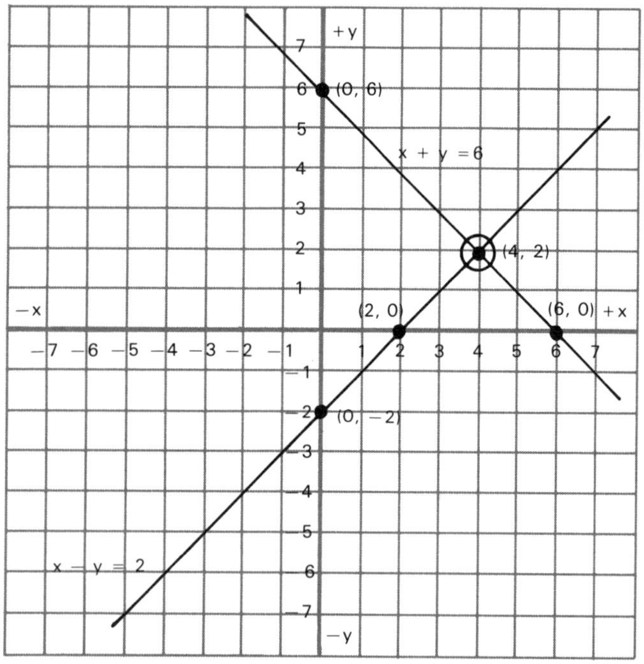

5. equation: 2x − y ≤ 0
   (0, 0) (3, 6) (1, 2): optional values
   equation: 3y − 2x ≤ 4
   $(0, 1\frac{1}{3})$ (−2, 0) (1, 2): optional values

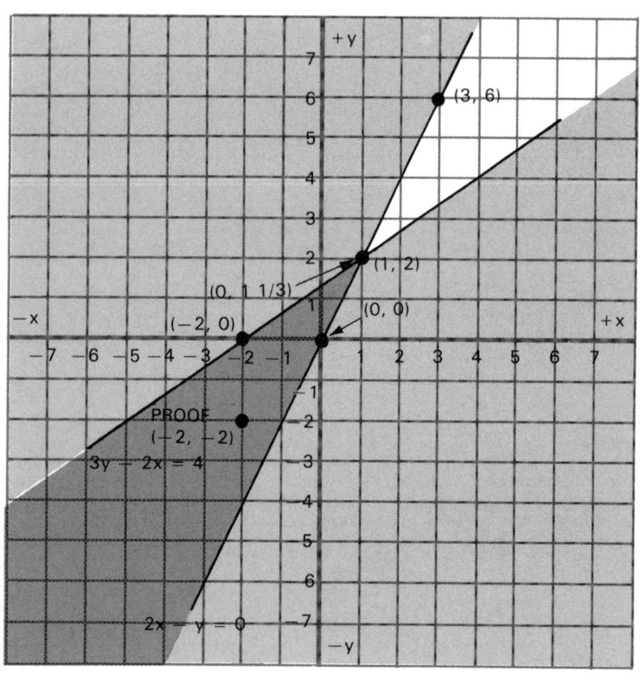

ANSWERS TO TEST MASTERS, page 151

1. x = 0, y = 0
2. linear, or first-degree
3. $\frac{1}{2}$

   $$m = \frac{y - y_1}{x - x_1} = \frac{4 - 1}{6 - 0} = \frac{3}{6} = \frac{1}{2}$$
4. $y = \frac{1}{2}x + 1$
5. (−6, −2)
6. Yes. For every value of x there is only one value of y.
7. 6
8. x > −2
9.

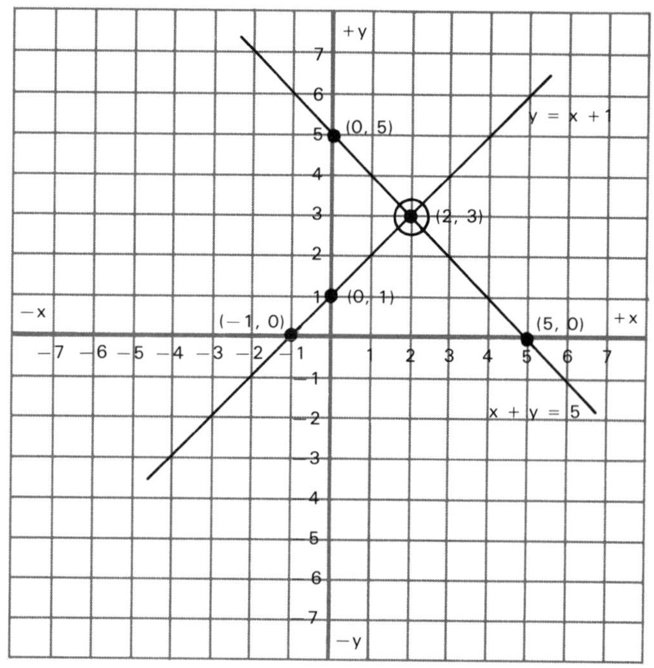

10. (2, 3)

Chapter 14
SOLVING PROBLEMS WITH ALGEBRA

Objectives

After studying this chapter, the student will be able to:

- Change word expressions into algebraic expressions.
- Solve word problems using algebra.
- Solve problems using ratio, proportion, or variation.

ANSWERS TO TEXT, page 221

Changing Word Expressions Into Algebraic Expressions

1. a. 2
   b. 9
   c. $\frac{x}{3}$
   d. 3(y + 4)
   e. $\frac{4z}{3}$

2. a. 200
   b. 160
   c. a − 40
   d. 3b + 40
   e. 4c − 39
3. a. 31
   b. 13
   c. 38 − x
   d. 38 − 2y
   e. 36 − z

Solving Word Problems

4. a. width = x
      length = 5x
   b. 168 = 2(5x) + 2(x)
      x = 14
   d. width = 14 ft.
      length = 70 ft.
5. a. side a = 3x + 2
      side b = 3x
      side c = x
   b. 37 = (3x + 2) + 3x + x
      x = 5
   d. side a = 17 in.
      side b = 15 in.
      side c = 5 in.
6. a. Melissa at present = x
      Mrs. Smith at present = 3x
      Melissa in 12 years = x + 12
      Mrs. Smith in 12 years = 3x + 12
   b. 2(x + 12) = 3x + 12
      x = 12
   d. Melissa at present = 12 yr.
      Mrs. Smith at present = 36 yr.
7. a. roofing nails at start = 2x
      drywall screws at start = x
      roofing nails at end = 2x − 2
      drywall screws at end = x − 2
   b. 2x − 2 = (x − 2) + 6
      x = 6
   d. roofing nails at start = 12 lb.
      drywall screws at start = 6 lb.
8. a. # of nickels = 3x
      # of dimes = x
      value of nickels = 5(3x)
      value of dimes = 10(x)
   b. 5(3x) + 10x = 125
      x = 5
   d. # of nickels = 15
      # of dimes = 5
9. a. # of pennies = 3x + 6
      # of nickels = x
      # of dimes = 3x
      value of pennies = 1(3x + 6)
      value of nickels = 5(x)
      value of dimes = 10(3x)
   b. (3x + 6) + 5(x) + 10(3x) = 234
      x = 6
   d. # of pennies = 24
      # of nickels = 6
      # of dimes = 18
10. a. number of hours = x
       distance traveled by 1st truck = 35x
       distance traveled by 2nd truck = 40x
    b. total miles = miles of 1st truck + miles of 2nd truck
       150 = 35x + 40x
       x = 2
    d. time trucks meet = 2 hr.
       distance from warehouse = 70 mi.
11. a. speed of car = x
       distance of car = 2x
       distance of bus = 5(30)
    b. 2x = 5(30)
       x = 75
    d. speed of car = 75 mph
12. a. 1st number = 4x
       2nd number = x
    b. 4x + x = 30
       x = 6
    d. 1st number = 24
       2nd number = 6
13. a. number = x
       one-half a number = $\frac{1}{2}x$
       one-third a number = $\frac{1}{3}x$
    b. $\frac{1}{2}x + \frac{1}{3}x = 5$
       x = 6
    d. number = 6
14. a. 1st integer = x
       2nd integer = x + 2
       3rd integer = x + 4
    b. x + (x + 2) + (x + 4) = 90
       x = 28
    d. 1st integer = 28
       2nd integer = 30
       3rd integer = 32
15. a. 1st integer = x
       2nd integer = x + 2
       3rd integer = x + 4
    b. x + (x + 2) = (x + 4) + 3
       x = 5
    d. 1st integer = 5
       2nd integer = 7
       3rd integer = 9
16. a. # of \$2 style = x
       # of \$1 style = 3000 − x
       value of \$2 style = 2(x)
       value of \$1 style = 1(3000 − x)
    b. 2x + (3000 − x) = 4850
       x = 1850
    d. # of \$2 style = 1850
       # of \$1 style = 1150
17. a. amount of red = x
       amount of white = 4x
    b. x + 4x = 70
       x = 14
    d. amount of red = 14 gal.
       amount of white = 56 gal.

18. a. water to be added = x
total amount in new mix = 20 + x
amount of acid = 30% of 20 = 6
b. 0.15(20 + x) = 6
x = 20
d. water to be added = 20 oz.
19. a. water to be added = x
total water in new solution = 20 + x
b. 0.08(20 + x) = 2
x = 5
d. water to be added = 5 cups
20. a. antifreeze to be added = x
antifreeze in starting mix = 20% of 25 = 5
antifreeze in new mix = 5 + x
total amount of new mix = 25 + x
b. 5 + x = 0.6(25 + x)
x = 25
d. antifreeze to be added = 25 qt.
21. a. insecticide to be added = x
insecticide in starting mix = 10% of 20 = 2
insecticide in new mix = 2 + x
total amount in new mix = 20 + x
b. 0.333(20 + x) = 2 + x
x = 7
d. insecticide to be added = 7 gal.
22. a. # of trips working together = x

working together
fraction of steel per trip = $\frac{1}{x}$

truck #1
fraction of steel per trip = $\frac{1}{24}$

truck #2
fraction of steel per trip = $\frac{1}{48}$

b. $\frac{1}{24} + \frac{1}{48} = \frac{1}{x}$

x = 16
d. # of trips working together = 16
23. a. time working together = x

working together
fraction plowed per hr. = $\frac{1}{x}$

large tractor
fraction plowed per hr. = $\frac{1}{12}$

small tractor
fraction plowed per hr. = $\frac{1}{36}$

b. $\frac{1}{12} + \frac{1}{36} = \frac{1}{x}$

x = 9
d. time working together = 9 hr.

Ratio, Proportion, and Variation

24. a. 2x + 5x = 49
x = 7
c. maximum developer = 14 oz.
maximum water = 35 oz.
25. a. $3\frac{1}{4}x = 13$

x = 4

height = $2\frac{1}{4}x$

c. height = 9 in.
26. a. 3x + x + 2x = 96
x = 16
c. orange juice = 48 oz.
pineapple juice = 16 oz.
ginger ale = 32 oz.

27. a. $\frac{x \text{ ft.}}{27 \text{ ft.}} = \frac{5 \text{ ft.}}{3 \text{ ft.}}$

x = 45
c. height of flagpole = 45 ft.

28. a. $\frac{152 \text{ mi.}}{4 \text{ hr.}} = \frac{247 \text{ mi.}}{x \text{ hr.}}$

x = 6.5
c. travel time = 6.5 hr.

29. a. $\frac{184 \text{ sample}}{6 \text{ defective}} = \frac{1288 \text{ sample}}{x \text{ defective}}$

x = 42
c. number defective = 42

30. a. $\frac{\frac{3}{4} \text{ in.}}{10 \text{ mi.}} = \frac{6 \text{ in.}}{x \text{ mi.}}$

x = 80
c. 6 in. = 80 mi.

31. P = kA
32. V = kr^3
33. F = ks
34. I = k/r^2
35. R = kL/A

ANSWERS TO WORKBOOK, page 133

Changing Word Expressions Into Algebraic Expressions

1. a. 3
b. 4
c. $\frac{x}{2}$
d. 6y
e. 2(z + 1)
2. a. 14
b. 16
c. a − 6
d. 4b + 6
e. c − 3
3. a. 15 yr.
b. 17 yr.
c. 28 − x yr.
d. 33 − y yr.
e. 30 − 2z yr.

Solving Word Problems

4. a. length = **x + 14**
   width = **x**
   b. **264 = 2(x + 14) + 2x**
   x = 59
   d. length = **73 ft.**
   width = **59 ft.**
5. a. angle A = **x + 31**
   angle B = **x**
   angle C = **x + 26**
   b. **(x + 31) + x + (x + 26) = 180**
   x = 41
   d. angle A = **72°**
   angle B = **41°**
   angle C = **67°**
6. a. red car at present = **2x**
   white car at present = **x**
   red car 4 years ago = **2x − 4**
   white car 4 years ago = **x − 4**
   b. **(2x − 4) = 4(x − 4)**
   x = 6
   d. red car at present = **12 yr.**
   white car at present = **6 yr.**
7. a. weight of crate = **30**
   weight of instrument = **x**
   b. **x + 30 = 150**
   x = 120
   d. weight of instrument = **120 lb.**
8. a. # of dimes = **x + 6**
   # of quarters = **x**
   value of dimes = **10(x + 6)**
   value of quarters = **25x**
   b. **10(x + 6) + 25x = 200**
   x = 4
   d. # of dimes = **10**
   # of quarters = **4**
9. a. # of quarters = **x**
   # of nickels = **16 − x**
   value of quarters = **25x**
   value of nickels = **5(16 − x)**
   b. **25x + 5(16 − x) = 140**
   x = 3
   d. # of quarters = **3**
   # of nickels = **13**
10. a. travel time = **x**
    distance traveled by 1st truck = **50x**
    distance traveled by 2nd truck = **55x**
    b. **50x + 55x = 735**
    x = 7
    d. travel time = **7 hr.**
11. a. travel time of 1st car = **x**
    travel time of 2nd car = **x − 2**

    speed of 2nd car = $\mathbf{\frac{180}{x - 2}}$

    b. **30x = 180**

    x = 6

    d. speed of 2nd car = **45 mph**
12. a. total value = **x**

    son's portion = $\mathbf{\frac{1}{4}x}$

    daughter's portion = $\mathbf{\frac{1}{3}x}$

    wife's portion = $\mathbf{\frac{1}{4}x + 4000}$

    b. $\mathbf{\frac{1}{4}x + \frac{1}{3}x + (\frac{1}{4}x + 4000) = x}$
    x = 24,000
    d. total value = **$24,000**
13. a. # in 1st pile = **x + 20**
    # in 2nd pile = **x**
    b. **x + (x + 20) = 140**
    x = 60
    d. # in 1st pile = **80**
    # in 2nd pile = **60**
14. a. 1st integer = **x**
    2nd integer = **x + 2**
    3rd integer = **x + 4**
    4th integer = **x + 6**
    b. **x + (x + 2) + (x + 4) + (x + 6) = −84**
    x = −24
    d. 1st integer = **−24**
    2nd integer = **−22**
    3rd integer = **−20**
    4th integer = **−18**
15. a. 1st number = **x**
    2nd number = **4x + 4**
    b. **x + (4x + 4) = 124**
    x = 24
    d. 1st number = **24**
    2nd number = **100**
16. a. total weight of 38¢ candy = **x**
    total weight of 70¢ candy = **80 − x**
    total value of 38¢ candy = **38x**
    total value of 70¢ candy = **70(80 − x)**
    b. **38x + 70(80 − x) = 50(80)**
    x = 50
    d. total weight of 38¢ candy = **50 lb.**
    total weight of 70¢ candy = **30 lb.**
17. a. # of small = **x**
    # of large = **5x**
    b. **x + 5x = 120**
    x = 20
    d. # of small = **20**
    # of large = **100**
18. a. butterfat-free milk to be added = **x**
    total amount in new mix = **400 + x**
    amount of butterfat = 5% of 400 = **20**
    b. **0.04(400 + x) = 20**
    x = 100
    d. butterfat-free milk to be added = **100 qt.**
19. a. water to be added = **x**
    total amount in new mix = **6 + x**
    amount of antifreeze = 33.3% of 6 = **2**
    b. **0.2(6 + x) = 2**
    x = 4
    d. water to be added = **4 gal.**

20. a. salt to be added = **x**
salt in starting mix = 25% of 32 = **8**
salt in new mix = **8 + x**
total amount in new mix = **32 + x**
b. **8 + x = 0.4(32 + x)**
x = 8
d. salt to be added = **8 lb.**
21. a. vinegar to be added = **x**
vinegar in starting mix = 10% of 40 = **4**
vinegar in new mix = **4 + x**
total amount in new mix = **40 + x**
b. **4 + x = 0.25(40 + x)**
x = 8
d. vinegar to be added = **8 oz.**
22. a. time working together = **x**

working together
\# of nails per # of sec. = $\mathbf{\frac{75}{x}}$

older model
\# of nails per # of sec. = $\mathbf{\frac{75}{60}}$

new model
\# of nails per # of sec. = $\mathbf{\frac{75}{30}}$

b. $\mathbf{\frac{75}{60} + \frac{75}{30} = \frac{75}{x}}$

x = 20
d. time working together = **20 sec.**
23. a. time working together = **x**

working together
fraction filled in 1 hr. = $\mathbf{\frac{1}{x}}$

fraction filled in 1 hr. = $\mathbf{\frac{1}{4}}$

fraction drained in 1 hr. = $\mathbf{\frac{1}{6}}$

b. $\mathbf{\frac{1}{4} - \frac{1}{6} = \frac{1}{x}}$

x = 12
d. time working together = **12 hr.**

Ratio, Proportion, and Variation

24. a. **7x + 6x = 260**
x = 20
c. women = **140**
men = **120**
25. a. **3x + 7x = 20**
x = 2
c. developer = **6 oz.**
water = **14 oz.**
26. a. **2x + 4x + 5x = 88**
x = 8
c. peanuts = **16 cups**
pretzels = **32 cups**
cereal = **40 cups**
27. a. $\mathbf{\frac{x\ ft.}{28\ ft.} = \frac{6\ ft.}{4\ ft.}}$
x = 42
c. height of tree = **42 ft.**
28. a. $\mathbf{\frac{2\ mg}{15\ lb.} = \frac{x\ mg}{60\ lb.}}$

x = 8
c. dosage of medicine = **8 mg**
29. a. $\mathbf{\frac{2\ computers}{5\ families} = \frac{x\ computers}{3580\ families}}$

x = 1432
c. families with computers = **1432**
30. a. $\mathbf{\frac{9\ cups}{3\ loaves} = \frac{x\ cups}{10\ loaves}}$

x = 30
c. amount of flour = **30 cups**
31. $F = k/d^2$
32. $I = kT^4$
33. $T = k\sqrt{l}$
34. $A = kr^2$
35. $L = kr^4/l^2$

ANSWERS TO TEST MASTERS, page 153

1. a. 50
b. 35
c. 125 − x
d. 125 − (y + 5)
e. 125 − 3z
2. 8 in.
Relationships: sides of square = x
length of rectangle = x + 12
width of rectangle = x + 4
Equation: 2(4x) = 2(x + 12) + 2(x + 4)
x = 8
Solution: sides of square = 8 in.
3. 9 hr.
Relationships: flight time of faster plane = x
flight time of slower plane = x + 3
distance of faster plane = 100x
distance of slower plane = 75(x + 3)
Equation: 75(x + 3) = 100x
x = 9
Solution: flight time of faster plane = 9 hr.
4. 28 ft., 12 ft.
Relationships: 1st piece = 2x + 4
2nd piece = x
Equation: (2x + 4) + x = 40
x = 12
Solution: 1st piece = 28 ft.
2nd piece = 12 ft.
5. 9, 11, 13
Relationships: 1st integer = x
2nd integer = x + 2
3rd integer = x + 4
Equation: x + 2(x + 2) + 3(x + 4) = 70
x = 9
Solution: 1st = 9
2nd = 11
3rd = 13
6. 72, 28
Relationships: bushels of 90¢ apples = x
bushels of 50¢ apples = 100 − x

value of 90¢ apples sold = 90x
value of 50¢ apples sold = 50(100 − x)
Equation: 90x + 50(100 − x) = 7880
x = 72
Solution: 90¢ bushels = 72
50¢ bushels = 28

7. 6 qt.
Relationships: antifreeze to be added = x
antifreeze in starting mix = 20% of 10 = 2
antifreeze in new mix = x + 2
total amount in new mix = 10 + x
Equation: 0.5(10 + x) = x + 2
x = 6
Solution: antifreeze to be added = 6 qt.

8. 2.25 hr., or 2 hr., 15 min.
Relationships:
combined operation: total time to print = x
fraction of checks printed in 1 hour:

combined operation $= \frac{1}{x}$

new computer $= \frac{1}{3}$

old computer $= \frac{1}{9}$

Equation: $\frac{1}{3} + \frac{1}{9} = \frac{1}{x}$
x = 2.25 hr.
Solution: combined operation: total time to print = 2.25 hr.

9. 16 oz. developer, 48 oz. water
Equation: 2x + 6x = 64
x = 8
Solution: developer = 16 oz.
water = 48 oz.

10. 60 ft.
Equation: $\frac{x \text{ ft.}}{90 \text{ ft.}} = \frac{4 \text{ ft.}}{6 \text{ ft.}}$
x = 60
Solution: height of building = 60 ft.

11. 25
General equation: y = kx
Evaluate for x = 8: 20 = k(8)
$k = \frac{5}{2}$

Evaluate for x = 10: $y = \frac{5}{2}(10)$
y = 25

12. 360
General equation: $y = k\left(\frac{x}{z}\right)$

Evaluate for x = 4, z = 10: $120 = k\left(\frac{4}{10}\right)$
k = 300

Evaluate for x = 6, z = 5: $y = 300\left(\frac{6}{5}\right)$
y = 360

Chapter 15
APPLIED TRIGONOMETRY

Objectives

After studying this chapter, the student will be able to:

- Explain basic terminology of trigonometry.
- Solve right triangles using the Pythagorean theorem and trigonometric functions.
- Solve oblique triangles using the law of sines and the law of cosines.
- Convert degrees to radians.
- Draw and label the unit circle.

ANSWERS TO TEXT, page 240

Types of Triangles

1. equilateral
2. isosceles
3. scalene
4. They are *similar triangles.*

Solving Right Triangles

5. Formula: $c = \sqrt{a^2 + b^2}$
Substitution: $c = \sqrt{(6 \text{ ft.})^2 + (12 \text{ ft.})^2}$
Solution: c = length of roof rafter = 13.4 ft.

6. a. Formula: $\tan A = \frac{\text{opposite}}{\text{adjacent}}$
Substitution: $\tan 35° = \frac{\text{opposite}}{100 \text{ ft.}}$
Solution: opposite = height of building = 70 ft.
b. Formula: $c = \sqrt{a^2 + b^2}$
Substitution: $c = \sqrt{(70 \text{ ft.})^2 + (100 \text{ ft.})^2}$
Solution: c = path of sunlight = 122.1 ft.
c. Formula: B = 90 − A
Substitution: B = 90° − 35°
Solution: B = 55°

7. a. Formula: $\cos A = \frac{\text{adjacent}}{\text{hypotenuse}}$
Substitution: $\cos 60° = \frac{8 \text{ ft.}}{\text{hypotenuse}}$
Solution: hypotenuse: = length of guy wire = 16 ft.
b. Formula: $\tan A = \frac{\text{opposite}}{\text{adjacent}}$
Substitution: $\tan 60° = \frac{\text{opposite}}{8 \text{ ft.}}$
Solution: opposite = height of antenna = 13.9 ft.
c. Formula: B = 90 − A
Substitution: B = 90° − 60°
Solution: B = 30°

8. a. Formula: $\tan B = \frac{\text{opposite}}{\text{adjacent}}$

Substitution: $\tan 70^\circ = \frac{\text{opposite}}{5 \text{ mi.}}$

Solution: opposite = length of run = 13.7 mi.

b. Formula: $c = \sqrt{a^2 + b^2}$
Substitution: $c = \sqrt{(5 \text{ mi.})^2 + (13.7 \text{ mi.})^2}$
Solution: c = length of road surface = 14.6 mi.

c. Formula: A = 90 − B
Substitution: A = 90° − 70°
Solution: A = 20°

9. a. Formula: $\sin A = \frac{\text{opposite}}{\text{hypotenuse}}$

Substitution: $\sin 45^\circ = \frac{300 \text{ yd.}}{\text{hypotenuse}}$

Solution: hypotenuse = travel distance = 424.3 yd.

b. Formula: $\tan A = \frac{\text{opposite}}{\text{adjacent}}$

Substitution: $\tan 45^\circ = \frac{300 \text{ yd.}}{\text{adjacent}}$

Solution: adjacent = width of river = 300 yd.

c. Formula: B = 90 − A
Substitution: B = 90° − 45°
Solution: B = 45°

10. a. Formula: $\cos A = \frac{\text{adjacent}}{\text{hypotenuse}}$

Substitution: $\cos 15^\circ = \frac{10 \text{ ft.}}{\text{hypotenuse}}$

Solution: hypotenuse = length of roof rafter = 10.4 ft.

b. Formula: $\tan A = \frac{\text{opposite}}{\text{adjacent}}$

Substitution: $\tan 15^\circ = \frac{\text{opposite}}{10 \text{ ft.}}$

Solution: opposite = height of attic = 2.7 ft.

c. Formula: B = 90 − A
Substitution: B = 90° − 15°
Solution: B = 75°

11. a. Formula: $c = \sqrt{a^2 + b^2}$
Substitution: $c = \sqrt{(20 \text{ ft.})^2 + (8 \text{ ft.})^2}$
Solution: c = length of ladder = 21.5 ft.

b. Formula: $\tan A = \frac{\text{opposite}}{\text{adjacent}}$

Substitution: $\tan A = \frac{20 \text{ ft.}}{8 \text{ ft.}}$

Solution: A = 68.2°

c. Formula: B = 90 − A
Substitution: B = 90° − 68.2°
Solution: B = 21.8°

12. a. Formula: $c = \sqrt{a^2 + b^2}$
Substitution: $c = \sqrt{(20 \text{ ft.})^2 + (24 \text{ ft.})^2}$
Solution: c = length of diagonal = 31.2 ft.

b. Formula: $\tan A = \frac{\text{opposite}}{\text{adjacent}}$

Substitution: $\tan A = \frac{20 \text{ ft.}}{24 \text{ ft.}}$

Solution: A = 39.8°

c. Formula: B = 90 − A
Substitution: B = 90° − 39.8°
Solution: B = 50.2°

13. a. Formula: $c = \sqrt{a^2 + b^2}$
Substitution: $c = \sqrt{(8 \text{ ft.})^2 + (12 \text{ ft.})^2}$
Solution: c = length of stairway stringer = 14.4 ft.

b. Formula: $\tan A = \frac{\text{opposite}}{\text{adjacent}}$

Substitution: $\tan A = \frac{8 \text{ ft.}}{12 \text{ ft.}}$

Solution: A = 33.7°

c. Formula: B = 90 − A
Substitution: B = 90° − 33.7°
Solution: B = 56.3°

14. a. Formula: $c = \sqrt{a^2 + b^2}$
Substitution: $c = \sqrt{(6 \text{ ft.})^2 + (6 \text{ ft.})^2}$
Solution: c = length of rafter a = 8.5 ft.

b. Formula: $c = \sqrt{a^2 + b^2}$
Substitution: $c = \sqrt{(6 \text{ ft.})^2 + (12 \text{ ft.})^2}$
Solution: c = length of rafter b = 13.4 ft.

c. Formula: $\tan A = \frac{\text{opposite}}{\text{adjacent}}$

Substitution: $\tan A = \frac{6 \text{ ft.}}{12 \text{ ft.}}$

Solution: A = 26.6°

15. a. Formula: $c = \sqrt{a^2 + b^2}$
Substitution: $c = \sqrt{(14.5 \text{ in.})^2 + (11 \text{ in.})^2}$
Solution: c = length of cross bridging = 18.2 in.

b. Formula: $\tan A = \frac{\text{opposite}}{\text{adjacent}}$

Substitution: $\tan A = \frac{14.5 \text{ in.}}{11 \text{ in.}}$

Solution: A = 52.8°

c. Formula: B = 90 − A
Substitution: B = 90° − 52.8°
Solution: B = 37.2°

16. a. Formula: $\sin A = \frac{\text{opposite}}{\text{hypotenuse}}$

Substitution: $\sin 30^\circ = \frac{\text{opposite}}{12 \text{ ft.}}$

Solution: opposite = height of vertical support = 6 ft.

b. Formula: $\cos A = \frac{\text{adjacent}}{\text{hypotenuse}}$

Substitution: $\cos 30° = \frac{\text{adjacent}}{12 \text{ ft.}}$

Solution: adjacent = length of base support = 10.4 ft.

c. Formula: $B = 90 - A$
Substitution: $B = 90° - 30°$
Solution: $B = 60°$

17. a. Formula: $\sin A = \frac{\text{opposite}}{\text{hypotenuse}}$

Substitution: $\sin 60° = \frac{\text{opposite}}{40 \text{ ft.}}$

Solution: opposite = height of wall = 34.6 ft.

b. Formula: $\cos A = \frac{\text{adjacent}}{\text{hypotenuse}}$

Substitution: $\cos 60° = \frac{\text{adjacent}}{40 \text{ ft.}}$

Solution: adjacent = length of base = 20 ft.

c. Formula: $B = 90 - A$
Substitution: $B = 90° - 60°$
Solution: $B = 30°$

18. a. Formula: $\cos A = \frac{\text{adjacent}}{\text{hypotenuse}}$

Substitution: $\cos 15° = \frac{\text{adjacent}}{10 \text{ mi.}}$

Solution: adjacent = length of run = 9.7 mi.

b. Formula: $\sin A = \frac{\text{opposite}}{\text{hypotenuse}}$

Substitution: $\sin 15° = \frac{\text{opposite}}{10 \text{ mi.}}$

Solution: opposite = height of rise = 2.6 mi.

c. Formula: $A = 90 - B$
Substitution: $A = 90° - 15°$
Solution: $A = 75°$

19. a. Formula: $\sin A = \frac{\text{opposite}}{\text{hypotenuse}}$

Substitution: $\sin 45° = \frac{\text{opposite}}{16 \text{ ft.}}$

Solution: opposite = length of run = 11.3 ft.

b. Formula: $\cos A = \frac{\text{adjacent}}{\text{hypotenuse}}$

Substitution: $\cos 45° = \frac{\text{adjacent}}{16 \text{ ft.}}$

Solution: adjacent = height of rise = 11.3 ft.

c. Formula: $B = 90 - A$
Substitution: $B = 90° - 45°$
Solution: $B = 45°$

20. a. Formula: $\sin A = \frac{\text{opposite}}{\text{hypotenuse}}$

Substitution: $\sin 20° = \frac{\text{opposite}}{32 \text{ ft.}}$

Solution: opposite = height of rise = 10.9 ft.

b. Formula: $c = \sqrt{a^2 + b^2}$
Substitution: $c = \sqrt{(10.9 \text{ ft.})^2 + (4 \text{ ft.})^2}$
Solution: c = length of long web = 11.6 ft.

c. Formula: $\cos A = \frac{\text{adjacent}}{\text{hypotenuse}}$

Substitution: $\cos 20° = \frac{\text{adjacent}}{32 \text{ ft.}}$

Solution: adjacent = 30.1 ft.
distance of 1/3 point = adjacent − 4 ft.
distance of 1/3 point = 30.1 ft. − 4 ft.
= 26.1 ft.

Solving Oblique Triangles

21. a. Formula: $B = 180 - (A + C)$
Substitution: $B = 180° - (30° + 110°)$
Solution: $B = 40°$

b. Formula: $\frac{a}{\sin A} = \frac{b}{\sin B}$

Substitution: $\frac{a}{\sin 30°} = \frac{350 \text{ ft.}}{\sin 40°}$

Solution: a = length of westerly boundary = 272.25 ft.

c. Formula: $\frac{c}{\sin C} = \frac{b}{\sin B}$

Substitution: $\frac{c}{\sin 110°} = \frac{350 \text{ ft.}}{\sin 40°}$

Solution: c = length of northerly boundary = 511.67 ft.

22. a. Formula: $B = 180 - (A + C)$
Substitution: $B = 180° - (60° + 80°)$
Solution: $B = 40°$

b. Formula: $\frac{a}{\sin A} = \frac{c}{\sin C}$

Substitution: $\frac{a}{\sin 60°} = \frac{28 \text{ ft.}}{\sin 80°}$

Solution: a = length of rafter a = 24.62 ft.

c. Formula: $\frac{b}{\sin B} = \frac{c}{\sin C}$

Substitution: $\frac{b}{\sin 40°} = \frac{28 \text{ ft.}}{\sin 80°}$

Solution: b = length of rafter b = 18.28 ft.

23. a. Formula: $\frac{a}{\sin A} = \frac{b}{\sin B}$

Substitution: $\frac{6 \text{ in.}}{\sin 60°} = \frac{6 \text{ in.}}{\sin B}$

Solution: $B = 60°$

b. Formula: $C = 180 - (A + B)$
Substitution: $C = 180° - (60° + 60°)$
Solution: $C = 60°$

c. Formula: $\frac{c}{\sin C} = \frac{a}{\sin A}$

Substitution: $\frac{c}{\sin 60°} = \frac{6 \text{ in.}}{\sin 60°}$

Solution: c = 6 in.

24. a. Formula: B = 180 − (A + C)
Substitution: B = 180° − (13° + 62°)
Solution: B = 105°

b. Formula: $\frac{c}{\sin C} = \frac{b}{\sin B}$

Substitution: $\frac{c}{\sin 62°} = \frac{40 \text{ ft.}}{\sin 105°}$

Solution: c = length of daredevil ramp = 36.56 ft.

25. a. Formula: $c = \sqrt{a^2 + b^2 - 2ab(\cos C)}$
Substitution: c =
$\sqrt{(7 \text{ ft.})^2 + (6 \text{ ft.})^2 - 2(7 \text{ ft.})(6 \text{ ft.})[\cos(180° - 65°)]}$

Solution: c = length of porch roof = 10.98 ft.

b. Formula: $\frac{a}{\sin A} = \frac{c}{\sin C}$

Substitution:
$\frac{7 \text{ ft.}}{\sin A} = \frac{10.98 \text{ ft.}}{\sin(180° - 65°)} = \frac{11 \text{ ft.}}{\sin 115°}$

Solution: A = 35.3°

c. Formula: B = 180 − (C + A)
Substitution: B = 180° − (115° + 35.3°)
Solution: B = 29.7°

26. a. Formula: $a = \sqrt{b^2 + c^2 - 2bc(\cos A)}$

Substitution: a =
$\sqrt{(3 \text{ in.})^2 + (4 \text{ in.})^2 - 2(3 \text{ in.})(4 \text{ in.})(\cos 45°)}$

Solution: a = 2.83 in.

b. Formula: $\frac{a}{\sin A} = \frac{b}{\sin B}$

Substitution: $\frac{2.83 \text{ in.}}{\sin 45°} = \frac{3 \text{ in.}}{\sin B}$

Solution: B = 48.55°

c. Formula: C = 180 − (A + B)
Substitution: C = 180° − (45° + 48.55°)
Solution: C = 86.45°

27. a. Formula: $A = \cos^{-1}\left(\frac{b^2 + c^2 - a^2}{2bc}\right)$

Substitution:
$A = \cos^{-1}\left(\frac{(100 \text{ ft.})^2 + (1 \text{ ft.})^2 - (100 \text{ ft.})^2}{2(100 \text{ ft.})(1 \text{ ft.})}\right)$

Solution: A = 89.71°

b. Formula: $\frac{a}{\sin A} = \frac{b}{\sin B}$

Substitution: $\frac{100 \text{ ft.}}{\sin 89.71°} = \frac{100 \text{ ft.}}{\sin B}$

Solution: B = 89.71°

c. Formula: C = 180 − (A + B)
Substitution: C = 180° − (89.71° + 89.71°)
Solution: C = 0.58°

28. a. Formula: $A = \cos^{-1}\left(\frac{b^2 + c^2 - a^2}{2bc}\right)$

Substitution:
$A = \cos^{-1}\left(\frac{(6 \text{ in.})^2 + (12 \text{ in.})^2 - (8 \text{ in.})^2}{2(6 \text{ in.})(12 \text{ in.})}\right)$

Solution: A = 36.34°

b. Formula: $\frac{a}{\sin A} = \frac{b}{\sin B}$

Substitution: $\frac{8 \text{ in.}}{\sin 36.34°} = \frac{6 \text{ in.}}{\sin B}$

Solution: B = 26.39°

c. Formula: C = 180 − (A + B)
Substitution: C = 180° − (36.34° + 26.39°)
Solution: C = 117.27°

ANSWERS TO WORKBOOK, page 151

Types of Triangles

1. equilateral
2. isosceles
3. scalene
4. yes

Solving Right Triangles

5. a. Formula: $\mathbf{c = \sqrt{a^2 + b^2}}$
Substitution: $\mathbf{c = \sqrt{(2 \text{ ft.})^2 + (6 \text{ ft.})^2}}$
Solution: c = **6.3 ft.**

6. a. Formula: $\mathbf{\sin A = \frac{opposite}{hypotenuse}}$

Substitution: $\mathbf{\sin 30° = \frac{6 \text{ ft.}}{hypotenuse}}$

Solution: hypotenuse = **12 ft.**

b. Formula: $\mathbf{\tan A = \frac{opposite}{adjacent}}$

Substitution: $\mathbf{\tan 30° = \frac{6 \text{ ft.}}{adjacent}}$

Solution: adjacent = **10.4 ft.**

c. Formula: **B = 90 − A**
Substitution: **B = 90° − 30°**
Solution: B = **60°**

7. a. Formula: $\mathbf{\tan A = \frac{opposite}{adjacent}}$

Substitution: $\mathbf{\tan 75° = \frac{opposite}{12 \text{ ft.}}}$

Solution: opposite = **44.8 ft.**

b. Formula: $\cos A = \mathbf{\frac{adjacent}{hypotenuse}}$

Substitution: $\mathbf{\cos 75° = \frac{12 \text{ ft.}}{hypotenuse}}$

Solution: hypotenuse = **46.4 ft.**

c. Formula: **B = 90 − A**
Substitution: **B = 90° − 75°**
Solution: B = **15°**

8. a. Formula: $\tan B = \frac{\text{opposite}}{\text{adjacent}}$

Substitution: $\tan 65° = \frac{\text{opposite}}{10 \text{ ft.}}$

Solution: opposite = **21.4 ft.**

b. Formula: $\cos B = \frac{\text{adjacent}}{\text{hypotenuse}}$

Substitution: $\cos 65° = \frac{10 \text{ ft.}}{\text{hypotenuse}}$

Solution: hypotenuse = **23.7 ft.**

c. Formula: $A = 90 - B$

Substitution: $A = 90° - 65°$

Solution: A = **25°**

9. a. Formula: $\sin A = \frac{\text{opposite}}{\text{hypotenuse}}$

Substitution: $\sin 45° = \frac{100 \text{ ft.}}{\text{hypotenuse}}$

Solution: hypotenuse = **141.4 ft.**

b. Formula: $\tan A = \frac{\text{opposite}}{\text{adjacent}}$

Substitution: $\tan 45° = \frac{100 \text{ ft.}}{\text{adjacent}}$

Solution: adjacent = **100 ft.**

c. Formula: $B = 90 - A$

Substitution: $B = 90° - 45°$

Solution: B = **45°**

10. a. Formula: $\cos A = \frac{\text{adjacent}}{\text{hypotenuse}}$

Substitution: $\cos 20° = \frac{16 \text{ ft.}}{\text{hypotenuse}}$

Solution: hypotenuse = **17 ft.**

b. Formula: $\tan A = \frac{\text{opposite}}{\text{adjacent}}$

Substitution: $\tan 20° = \frac{\text{opposite}}{16 \text{ ft.}}$

Solution: opposite = **5.8 ft.**

c. Formula: $B = 90 - A$

Substitution: $B = 90° - 20°$

Solution: B = **70°**

11. a. Formula: $c = \sqrt{a^2 + b^2}$

Substitution: $c = \sqrt{(12 \text{ ft.})^2 + (30 \text{ ft.})^2}$

Solution: c = **32.3 ft.**

b. Formula: $\tan A = \frac{\text{opposite}}{\text{adjacent}}$

Substitution: $\tan A = \frac{30 \text{ ft.}}{12 \text{ ft.}}$

Solution: A = **68.2°**

c. Formula: $B = 90 - A$

Substitution: $B = 90° - 68.2°$

Solution: B = **21.8°**

12. a. Formula: $c = \sqrt{a^2 + b^2}$

Substitution: $c = \sqrt{(32 \text{ ft.})^2 + (48 \text{ ft.})^2}$

Solution: c = **57.7 ft.**

b. Formula: $\tan A = \frac{\text{opposite}}{\text{adjacent}}$

Substitution: $\tan A = \frac{48 \text{ ft.}}{32 \text{ ft.}}$

Solution: A = **56.3°**

c. Formula: $B = 90 - A$

Substitution: $B = 90° - 56.3°$

Solution: B = **33.7°**

13. a. Formula: $c = \sqrt{a^2 + b^2}$

Substitution: $c = \sqrt{(9 \text{ ft.})^2 + (14 \text{ ft.})^2}$

Solution: c = **16.6 ft.**

b. Formula: $\tan A = \frac{\text{opposite}}{\text{adjacent}}$

Substitution: $\tan A = \frac{9 \text{ ft.}}{14 \text{ ft.}}$

Solution: A = **32.7°**

c. Formula: $B = 90 - A$

Substitution: $B = 90° - 32.7°$

Solution: B = **57.3°**

14. a. Formula: $c = \sqrt{a^2 + b^2}$

Substitution: $c = \sqrt{(8 \text{ ft.})^2 + (10 \text{ ft.})^2}$

Solution: c = **12.8 ft.**

b. Formula: $c = \sqrt{a^2 + b^2}$

Substitution: $c = \sqrt{(8 \text{ ft.})^2 + (18 \text{ ft.})^2}$

Solution: c = **19.7 ft.**

c. Formula: $\tan A = \frac{\text{opposite}}{\text{adjacent}}$

Substitution: $\tan A = \frac{8 \text{ ft.}}{18 \text{ ft.}}$

Solution: A = **24°**

15. a. Formula: $c = \sqrt{a^2 + b^2}$

Substitution: $c = \sqrt{(11 \text{ in.})^2 + (22.5 \text{ in.})^2}$

Solution: c = **25 in.**

b. Formula: $\tan A = \frac{\text{opposite}}{\text{adjacent}}$

Substitution: $\tan A = \frac{22.5 \text{ in.}}{11 \text{ in.}}$

Solution: A = **63.9°**

c. Formula: $B = 90 - A$

Substitution: $B = 90° - 63.9°$

Solution: B = **26.1°**

16. a. Formula: $\cos A = \frac{\text{adjacent}}{\text{hypotenuse}}$

Substitution: $\cos 35° = \frac{\text{adjacent}}{16 \text{ ft.}}$

Solution: adjacent = **13.1 ft.**

b. Formula: $\sin A = \frac{\text{opposite}}{\text{hypotenuse}}$

Substitution: $\sin 35° = \frac{\text{opposite}}{16 \text{ ft.}}$

Solution: opposite = **9.2 ft.**

c. Formula: $B = 90 - A$
Substitution: $B = 90° - 35°$
Solution: B = **55°**

17. a. Formula: $\sin A = \frac{\text{opposite}}{\text{hypotenuse}}$

Substitution: $\sin 65° = \frac{\text{opposite}}{32 \text{ ft.}}$

Solution: opposite = **29.0 ft.**

b. Formula: $\cos A = \frac{\text{adjacent}}{\text{hypotenuse}}$

Substitution: $\cos 65° = \frac{\text{adjacent}}{32 \text{ ft.}}$

Solution: adjacent = **13.5 ft.**

c. Formula: $B = 90 - A$
Substitution: $B = 90° - 65°$
Solution: B = **25°**

18. a. Formula: $\cos A = \frac{\text{adjacent}}{\text{hypotenuse}}$

Substitution: $\cos 20° = \frac{\text{adjacent}}{3 \text{ m}}$

Solution: adjacent = **2.8 m**

b. Formula: $\sin A = \frac{\text{opposite}}{\text{hypotenuse}}$

Substitution: $\sin 20° = \frac{\text{opposite}}{3 \text{ m}}$

Solution: opposite = **1 m**

c. Formula: $B = 90 - A$
Substitution: $B = 90° - 20°$
Solution: B = **70°**

19. a. Formula: $\sin A = \frac{\text{opposite}}{\text{hypotenuse}}$

Substitution: $\sin 40° = \frac{\text{opposite}}{12 \text{ ft.}}$

Solution: opposite = **7.7 ft.**

b. Formula: $\cos A = \frac{\text{adjacent}}{\text{hypotenuse}}$

Substitution: $\cos 40° = \frac{\text{adjacent}}{12 \text{ ft.}}$

Solution: adjacent = **9.2 ft.**

c. Formula: $B = 90 - A$
Substitution: $B = 90° - 40°$
Solution: B = **50°**

20. a. Formula: $\sin A = \frac{\text{opposite}}{\text{hypotenuse}}$

Substitution: $\sin 20° = \frac{\text{opposite}}{30 \text{ ft.}}$

Solution: opposite = **10.3 ft.**

b. Formula: $c = \sqrt{a^2 + b^2}$
Substitution: $c = \sqrt{(6 \text{ ft.})^2 + (10.3 \text{ ft.})^2}$
Solution: c = **11.9 ft.**

c. Formula: $\cos A = \frac{\text{adjacent}}{\text{hypotenuse}}$

Substitution: $\cos 20° = \frac{\text{adjacent}}{30 \text{ ft.}}$

Solution: adjacent = 28.2 ft.
distance of 1/3 point = adjacent − 6 ft.
= 28.2 ft. − 6 ft.
= **22.2 ft.**

Solving Oblique Triangles

21. a. Formula: $B = 180 - (A + C)$
Substitution: $B = 180° - (45° + 100°)$
Solution: $B = 35°$

b. Formula: $\frac{a}{\sin A} = \frac{b}{\sin B}$

Substitution: $\frac{a}{\sin 45°} = \frac{100 \text{ m}}{\sin 35°}$

Solution: **a = length of westerly boundary = 123.28 m**

c. Formula: $\frac{c}{\sin C} = \frac{b}{\sin B}$

Substitution: $\frac{c}{\sin 100°} = \frac{100 \text{ m}}{\sin 35°}$

Solution: **c = length of northerly boundary =171.7 m**

22. a. Formula: $B = 180 - (A + C)$
Substitution: $B = 180° - (65° + 85°)$
Solution: $B = 30°$

b. Formula: $\frac{a}{\sin A} = \frac{c}{\sin C}$

Substitution: $\frac{a}{\sin 65°} = \frac{26 \text{ ft.}}{\sin 85°}$

Solution: **a = length of rafter a = 23.65 ft.**

c. Formula: $\frac{b}{\sin B} = \frac{c}{\sin C}$

Substitution: $\frac{b}{\sin 30°} = \frac{26 \text{ ft.}}{\sin 85°}$

Solution: **b = length of rafter b = 13.05 ft.**

23. a. Formula: $\frac{a}{\sin A} = \frac{b}{\sin B}$

Substitution: $\frac{80 \text{ cm}}{\sin 60°} = \frac{80 \text{ cm}}{\sin B}$

Solution: $B = 60°$

b. Formula: $C = 180 - (A + B)$
Substitution: $C = 180° - (60° + 60°)$
Solution: $C = 60°$

c. Formula: $\frac{c}{\sin C} = \frac{a}{\sin A}$

Substitution: $\frac{c}{\sin 60°} = \frac{80 \text{ cm}}{\sin 60°}$

Solution: $c = 80$ cm

24. a. Formula: $B = 180 - (A + C)$
Substitution: $B = 180° - (21° + 100°)$
Solution: $B = 59°$

b. Formula: $\frac{c}{\sin C} = \frac{a}{\sin A}$

Substitution: $\frac{c}{\sin 100°} = \frac{5 \text{ ft.}}{\sin 21°}$

Solution: $c = 13.74$ ft.

25. a. Formula: $c = \sqrt{a^2 + b^2 - 2ab(\cos C)}$
Substitution:
$c = \sqrt{(15 \text{ ft.})^2 + (10 \text{ ft.})^2 - 2(15 \text{ ft.})(10 \text{ ft.})[\cos(180° - 70.5°)]}$
Solution: c = length of top chord = 20.62 ft.

b. Formula: $\frac{a}{\sin A} = \frac{c}{\sin C}$

Substitution:
$\frac{15 \text{ ft.}}{\sin A} = \frac{20.62 \text{ ft.}}{\sin(180° - 70.5°)} = \frac{20.62 \text{ ft.}}{\sin 109.5°}$

Solution: $A = 43.29°$
c. Formula: $B = 180 - (C + A)$
Substitution: $B = 180° - (109.5° + 43.29°)$
Solution: $B = 27.21°$

26. a. Formula: $a = \sqrt{b^2 + c^2 - 2bc(\cos A)}$
Substitution:
$a = \sqrt{(5 \text{ in.})^2 + (4 \text{ in.})^2 - 2(5 \text{ in.})(4 \text{ in.})(\cos 45°)}$

Solution: $a = 3.57$ in.

b. Formula: $\frac{a}{\sin A} = \frac{b}{\sin B}$

Substitution: $\frac{3.57 \text{ in.}}{\sin 45°} = \frac{5 \text{ in.}}{\sin B}$

Solution: $B = 82.03°$
c. Formula: $C = 180 - (A + B)$
Substitution: $C = 180 - (45 + 82.03°)$
Solution: $C = 52.97°$

27. a. Formula: $A = \cos^{-1}\left(\frac{b^2 + c^2 - a^2}{2bc}\right)$

Substitution:
$A = \cos^{-1}\left(\frac{(6\text{in.})^2 + (8 \text{ in.})^2 - (6 \text{ in.})^2}{2(6 \text{ in.})(8 \text{ in.})}\right)$

Solution: $A = 48.19°$

b. Formula: $\frac{a}{\sin A} = \frac{b}{\sin B}$

Substitution: $\frac{6 \text{ in.}}{\sin 48.19°} = \frac{6 \text{ in.}}{\sin B}$

Solution: $B = 48.19°$
c. Formula: $C = 180 - (A + B)$
Substitution: $C = 180° - (48.19° + 48.19°)$
Solution: $C = 83.62°$

28. a. Formula: $A = \cos^{-1}\left(\frac{b^2 + c^2 - a^2}{2\ bc}\right)$

Substitution:
$A = \cos^{-1}\left(\frac{(9 \text{ in.})^2 + (18 \text{ in.})^2 - (12 \text{ in.})^2}{2(9 \text{ in.})(18 \text{ in.})}\right)$

Solution: $A = 36.34°$

b. Formula: $\frac{a}{\sin A} = \frac{b}{\sin B}$

Substitution: $\frac{12 \text{ in.}}{\sin 36.34°} = \frac{9 \text{ in.}}{\sin B}$

Solution: $B = 26.39°$
c. Formula: $C = 180 - (A + B)$
Substitution: $C = 180° - (36.34° + 26.39°)$
Solution: $C = 117.27°$

ANSWERS TO TEST MASTERS, page 159

1. 5 ft., 6.7 ft.
   a. Formula: $c = \sqrt{a^2 + b^2}$
   Substitution: $c = \sqrt{(3 \text{ ft.})^2 + (4 \text{ ft.})^2}$
   Solution: c = 5 ft.
   b. Formula: $c = \sqrt{a^2 + b^2}$
   Substitution: $c = \sqrt{(3 \text{ ft.})^2 + (6 \text{ ft.})^2}$
   Solution: c = 6.7 ft.
2. 17.3 ft., 20 ft., 30°

   a. Formula: $\tan A = \frac{\text{opposite}}{\text{adjacent}}$

   Substitution: $\tan 60° = \frac{\text{opposite}}{10 \text{ ft.}}$

   Solution: opposite = 17.3 ft.

   b. Formula: $\cos A = \frac{\text{adjacent}}{\text{hypotenuse}}$

   Substitution: $\cos 60° = \frac{10 \text{ ft.}}{\text{hypotenuse}}$

   Solution: hypotenuse = 20 ft.
   c. Formula: $B = 90 - A$
   Substitution: $B = 90° - 60°$
   Solution: $B = 30°$

3. 20 ft., 36.9°, 53.1°
   a. Formula: $c = \sqrt{a^2 + b^2}$
   Substitution: $c = \sqrt{(12\text{ ft.})^2 + (16\text{ ft.})^2}$
   Solution: $c = 20$ ft.
   b. Formula: $\tan A = \frac{\text{opposite}}{\text{adjacent}}$
   Substitution: $\tan A = \frac{12\text{ ft.}}{16\text{ ft.}}$
   Solution: $A = 36.9°$
   c. Formula: $B = 90 - A$
   Substitution: $B = 90° - 36.9°$
   Solution: $B = 53.1°$

4. 11.3 ft., 11.3 ft., 45°
   a. Formula: $\sin A = \frac{\text{opposite}}{\text{hypotenuse}}$
   Substitution: $\sin 45° = \frac{\text{opposite}}{16\text{ ft.}}$
   Solution: opposite = 11.3 ft.
   b. Formula: $\cos A = \frac{\text{adjacent}}{\text{hypotenuse}}$
   Substitution: $\cos 45° = \frac{\text{adjacent}}{16\text{ ft.}}$
   Solution: adjacent = 11.3 ft.
   c. Formula: $B = 90 - A$
   Substitution: $B = 90° - 45°$
   Solution: $B = 45°$

5. 40°, 766 ft., 500 ft.
   a. Formula: $B = 180 - (A + C)$
   Substitution: $B = 180° - (40° + 100°)$
   Solution: $B = 40°$
   b. Formula: $\frac{c}{\sin C} = \frac{b}{\sin B}$
   Substitution: $\frac{c}{\sin 100°} = \frac{500\text{ ft.}}{\sin 40°}$
   Solution: c = length of northerly boundary = 766 ft.
   c. Formula: $\frac{a}{\sin A} = \frac{b}{\sin B}$
   Substitution: $\frac{a}{\sin 40°} = \frac{500\text{ ft.}}{\sin 40°}$
   Solution: a = length of westerly boundary = 500 ft.

6. 30°, 161.26″
   a. Formula: $B = 180 - (A + C)$
   Substitution: $B = 180° - (20° + 130°)$
   Solution: $B = 30°$
   b. Formula: $\frac{c}{\sin C} = \frac{a}{\sin A}$
   Substitution: $\frac{c}{\sin 130°} = \frac{72''}{\sin 20°}$
   Solution: c = length of base = 161.26″

7. 4.84″, 38.3°, 111.7°
   a. Formula: $a = \sqrt{b^2 + c^2 - 2bc(\cos A)}$
   Substitution:
   $a = \sqrt{(6\text{ in.})^2 + (9\text{ in.})^2 - 2(6\text{ in.})(9\text{ in.})(\cos 30°)}$
   Solution: a = measure of length a = 4.84″
   b. Formula: $\frac{b}{\sin B} = \frac{a}{\sin A}$
   Substitution: $\frac{6\text{ in.}}{\sin B} = \frac{4.84\text{ in.}}{\sin 30°}$
   Solution: $B = 38.3°$
   c. Formula: $C = 180 - (A + B)$
   Substitution: $C = 180 - (30° + 38.3°)$
   Solution: $C = 111.7°$

8. 86.42°, 86.42°, 7.16°
   a. Formula: $A = \cos^{-1}\left(\frac{b^2 + c^2 - a^2}{2\,bc}\right)$
   Substitution:
   $A = \cos^{-1}\left(\frac{(40\text{ cm})^2 + (5\text{ cm})^2 - (40\text{ cm})^2}{2(40\text{ cm})(5\text{ cm})}\right)$
   Solution: $A = 86.42°$
   b. Formula: $\frac{b}{\sin B} = \frac{a}{\sin A}$
   Substitution: $\frac{40}{\sin B} = \frac{40}{\sin 86.42°}$
   Solution: $B = 86.42°$
   c. Formula: $C = 180 - (A + B)$
   Substitution: $C = 180° - (86.42° + 86.42°)$
   Solution: $C = 7.16°$

Chapter 16
SCIENTIFIC AND ENGINEERING NOTATION

Objectives

After studying this chapter, the student will be able to:

- Convert numbers between ordinary, scientific, E, and engineering notation.
- Add, subtract, multiply, and divide power of 10 values.

ANSWERS TO TEXT, page 256

Scientific Notation

1. 8.9×10^5
2. 1.02×10^3
3. 3.25×10^2
4. 6.7×10^{-3}
5. 1.25×10^{-2}
6. 9.2×10^{-1}
7. 3.56×10^1
8. 2.98×10^2
9. 1.38×10^0
10. 2.7×10^6

Changing Improper Scientific Notation

11. 8.9×10^7
12. 3.56×10^1

13. 1.25×10^4
14. 9.7×10^0
15. 6.75×10^3
16. 3.2×10^{-1}
17. 2.58×10^{-6}
18. 1.5×10^1
19. 6.01×10^{-6}
20. 4.75×10^0

Changing Scientific Notation to Ordinary Notation

21. 7600
22. 0.0092
23. 25,800,000
24. 0.00000046
25. 0.0607
26. 0.013
27. 653
28. 0.0000211
29. 4.25
30. 10.9

E Notation

31. 2.5 E03
32. 3.5 E-03
33. 1.67 E05
34. 8.09 E-02
35. 1.0 E01
36. 1.5 E-06
37. 1.25 E03
38. 3.03 E-03
39. 6.7 E08
40. 5.0 E-03

Changing E Notation to Ordinary Notation

41. 3,560,000
42. 0.000021
43. 0.00000000067
44. 49
45. 12.2
46. 0.000045
47. 0.000807
48. 250,000
49. 0.0065
50. 4.8

Engineering Notation

51. 3500 kV, 3.5 MV
52. 0.00678 MW, 6.78 kW
53. 0.004 kHz, 4000 mHz
54. 5.8 mA, 5800 μA
55. 9 μF, 9000 nF
56. 250,000 W, 0.25 MW
57. 108,000 μV, 0.108 V
58. 7.5 kHz, 0.0000075 GHz
59. 13 μs, 0.000013 s
60. 0.045 nH, 45 pH

Arithmetic with Powers of 10

61. 7.57×10^3
62. 2.679×10^{-1}
63. 1.93 E04
64. 9.045 E00
65. 1.075 milli
66. 92.64 mega
67. 5.64×10^0
68. 1.847×10^0
69. 2.97 E-05
70. 1.25 E05
71. 87.8 mega
72. 0.401 pico
73. 9.36×10^7
74. 4.5×10^{-8}
75. 3.9 E01
76. 2.35 E03
77. 1.403 giga
78. 0.529 micro
79. 4.0×10^0
80. 3.0×10^2
81. 3.1 E-06
82. 1.37 E-02
83. 12
84. 80 milli

Word Problems with Scientific Notation

85. 1.5×10^{-4} sec.

$$100\ \cancel{\text{operations}}\left(\frac{1.5 \times 10^{-6}\ \text{sec.}}{\cancel{\text{operation}}}\right) = 150 \times 10^{-6}\ \text{sec.}$$

86. 7.9×10^3 mi. diameter
8.64×10^5 mi. $\times$ 0.9909 = 8.561×10^5 mi. smaller
8.64×10^5 mi. $-$ 8.561×10^5 mi.
$= 0.079 \times 10^5$ mi.

87. 10.0011 ft. pipe

$$\left(\frac{1.1 \times 10^{-5}\ \text{ft.}}{\cancel{\text{ft.}}\text{-}\cancel{^\circ\text{C}}}\right)\left(42\,\cancel{^\circ\text{C}} - 32\,\cancel{^\circ\text{C}}\right)\left(10\ \cancel{\text{ft.}}\right)$$
$$= 110 \times 10^{-5}\ \text{ft. expansion}$$

10 ft. + 1.1×10^{-3} ft. = 10 ft. + 0.0011 ft.
= 10.0011 ft.

88. 3.152×10^{-4} in. particle diameter

$$8\ \cancel{\text{microns}}\left(\frac{0.001\ \cancel{\text{mm}}}{\cancel{\text{micron}}}\right)\left(\frac{1\ \text{cm}}{10\ \cancel{\text{mm}}}\right) = 8.04 \times 10^{-4}\ \text{cm}$$

$8.04 \times 10^{-4} \times 0.394 = 3.152 \times 10^{-4}$ in.

89. 20,000 gal. tank

$$4.62 \times 10^6\ \cancel{\text{in.}^3}\left(\frac{1\ \cancel{\text{ft.}^3}}{1728\ \cancel{\text{in.}^3}}\right)\left(\frac{7.48052\ \text{gal.}}{\cancel{\text{ft.}^3}}\right) = 20,000\ \text{gal.}$$

90. 2.11×10^2 ft. diameter

$$2.5 \times 10^5\ \cancel{\text{bbl.}}\left(\frac{42\ \cancel{\text{gal.}}}{\cancel{\text{bbl.}}}\right)\left(\frac{1\ \text{ft.}^3}{7.48052\ \cancel{\text{gal.}}}\right) = 1,403,645.7\ \text{ft.}^3$$

1,403,645.7 ft.^{3} = A(40 ft.) A = 35,091.143 ft.^{2}

$$35,091.143\ \text{ft.}^2 = \frac{\pi d^2}{4}$$

d^2 = 44,679.431 ft.^{2} d = 211.4 ft.

91. 2.753×10^{-1} mi.

$$1454\ \cancel{\text{ft.}}\left(\frac{1\ \text{mi.}}{5280\ \cancel{\text{ft.}}}\right) = 0.2753\ \text{mi.}$$

92. 3,852,000 mi.^{2}
8.649×10^6 mi.^{2} $-$ 4.797×10^6 mi.^{2}
$= 3.852 \times 10^6$ mi.^{2}

93. lunar: 9.324×10^7 mi.; solar: 9.276×10^7 mi.
lunar eclipse: 9.3×10^7 mi. + 2.4×10^5 mi. $= 9.324 \times 10^7$ mi.
solar eclipse: 9.3×10^7 mi. − 2.4×10^5 mi. $= 9.276 \times 10^7$ mi.
94. 6.1976×10^3 m high
$2.032 \times 10^4 \times 0.305 = 6.1976 \times 10^3$ m
95. 1.248×10^{16} electrons

$$0.002\,\not{A}\left(\frac{6.24 \times 10^{18}\text{ electrons}}{1\,\not{A}}\right) = 0.01248 \times 10^{18}\text{ electrons}$$

ANSWERS TO WORKBOOK, page 165

Scientific Notation
1. 2.54×10^6
2. 6.82×10^3
3. 8.97×10^5
4. 3.01×10^2
5. 1.06×10^1
6. 1.25×10^{-1}
7. 3.6×10^{-4}
8. 9.1×10^{-3}
9. 5.7×10^0
10. 5.21×10^{-2}

Changing Improper Scientific Notation
11. 1.2×10^6
12. 4.5×10^3
13. 3.68×10^{-1}
14. 9.8×10^{-7}
15. 4.67×10^8
16. 8.75×10^{-7}
17. 4.5×10^1
18. 2.39×10^0
19. 7.27×10^0
20. 8.4×10^{-4}

Changing Scientific Notation to Ordinary Notation
21. 1560
22. 0.00745
23. 27,800
24. 0.00382
25. 670
26. 0.00000042
27. 3.12
28. 0.0025
29. 467
30. 0.251

E Notation
31. 4.5 E08
32. 2.02 E03
33. 3 E00
34. 6.1 E-02
35. 1.06 E-05
36. 5.54 E06
37. 3.42 E00
38. 4.8 E-06
39. 1.68 E-04
40. 3.2 E-03

Changing E Notation to Ordinary Notation
41. 607,000
42. 0.0072
43. 1930
44. 0.00000202
45. 4500
46. 0.000000081
47. 3.56
48. 47.9
49. 8.9
50. 0.0001

Engineering Notation
51.	**890** kW	**0.890** MW
52.	**0.00528** MV	**5.28** kV
53.	**0.01** kA	**10,000** mA
54.	**350** mΩ	**0.00035** kΩ
55.	**87,300** μs	**0.0873** s
56.	**0.235** MHz	**235,000** Hz
57.	**6.45** mH	**6450** μH
58.	**0.00047** mV	**470** nV
59.	**36.2** kW	**36,200** W
60.	**5000** mA	**0.005** kA

Arithmetic With Powers of 10
61. 5.978×10^3
62. 8.81×10^{-6}
63. 1.443 E05
64. 3.5 E-04
65. 39.6 micro
66. 94.19 mega
67. 8.495×10^5
68. 1.502×10^{-4}
69. 8.9 E-07
70. 2.125 E04
71. 52.44 pico
72. 430.3 mega
73. 4.65×10^{-1}
74. 1.922×10^6
75. 2.16 E11
76. 1.25 E-06
77. 150 kilo
78. 4.2 nano
79. 6.0×10^{-1}
80. 8.0×10^{-2}
81. 3.0 E-07
82. 6.0 E04
83. 20 micro
84. 6.8

ANSWERS TO TEST MASTERS, page 163
1. 1.28×10^6
2. 5.35×10^2
3. 4.0×10^0
4. 6.6×10^{-1}
5. 9.1×10^{-4}
6. 2.5×10^5
7. 7.5×10^0
8. 3.2×10^2
9. 5.6×10^{-2}
10. 9.4×10^{-6}
11. 8,900,000
12. 0.0047

13. 280
14. 0.00064
15. 601,000
16. 2.7 E11
17. 2.35 E-10
18. 6.18 E02
19. 1.05 E-02
20. 9.1 E00
21. 111,000,000
22. 0.0125
23. 460
24. 0.0000121
25. 0.57
26. **34** kW — **0.034** MW
27. **0.00125** kV — **1,250** mV
28. **460** μs — **0.00046** s
29. **0.630** MHz — **630,000** Hz
30. **50** kΩ — **50,000** Ω
31. 1.285×10^5
32. 4.299×10^{-6}
33. 1.023 E03
34. 50.65 milli
35. 114.8 giga
36. 3.49×10^{-2}
37. 3.94 E-04
38. 2.11 E07
39. 5.365 mega
40. 7.254 milli
41. 1.25×10^3
42. 4.6×10^{-8}
43. 7.44 E-03
44. 2.58 E01
45. 7
46. 2.0×10^4
47. 5.0×10^0
48. 6.0 E-11
49. 1.34 E11
50. 500

Chapter 17
BASIC RELATIONSHIPS OF ELECTRICITY

Objectives

After studying this chapter, the student will be able to:

- Explain basic relationships of electricity, including Ohm's Law, Kirchhoff's laws, and power formulas.
- Analyze series, parallel, and combination circuits using these laws.

ANSWERS TO TEXT, page 278

Ohm's Law/Watt's Law

1. V = 30 V, P = 450 W
2. V = 2.5 V, R = 1.25 Ω
3. R = 10 kΩ, P = 1 W
4. I = 50 A, P = 12.5 kW
5. I = 6.67 mA, R = 2.25 kΩ
6. V = 3.16 V, I = 3.16 mA
7. V = 250 V, P = 6.25 W
8. V = 12.5 V, R = 6.25 kΩ
9. R = 10 kΩ, P = 0.25 W
10. I = 2 mA, P = 10 mW
11. I = 0 A, P = 0 W
12. I = 1 A, R = 10 Ω
13. V = 0.1 V, P = 1 μW
14. V = 10 V, R = 100 Ω
15. R = 25 Ω, P = 25 W
16. I = 0.1 A, P = 2 W
17. I = 1 A, R = 1 Ω
18. V = 10 V, I = 10 mA
19. V = 10 kV, P = 10 kW
20. R = 100 Ω, P = 1 W

Series Circuits/Parallel Circuits

21. a. R_T = 600 Ω
    b. I_T = 0.1 A
    c. V_{R_1} = 20 V
    d. V_{R_2} = 15 V
    e. V_{R_3} = 25 V
    f. P_{R_1} = 2 W
    g. P_{R_2} = 1.5 W
    h. P_{R_3} = 2.5 W
    i. P_T = 6 W
22. a. R_T = 100 Ω
    b. I_T = 0.5 A
    c. V_{R_1} = 5 V
    d. V_{R_3} = 20 V
    e. V_T = 50 V
    f. P_{R_1} = 2.5 W
    g. P_{R_2} = 12.5 W
    h. P_{R_3} = 10 W
    i. P_T = 25 W
23. a. R_T = 6 kΩ
    b. R_3 = 600 Ω
    c. P_T = 5.4 W
24. a. R_1 = 60 Ω
    b. R_T = 120 Ω
    c. V_T = 30 V
    d. V_{R_2} = 10 V
    e. V_{R_3} = 5 V
    f. P_T = 7.5 W
25. a. I_T = 20 mA
    b. R_T = 600 Ω
    c. R_2 = 80 Ω
    d. P_{R_1} = 160 mW
    e. P_{R_2} = 32 mW
    f. P_T = 240 mW
26. a. R_T = 34.3 Ω
    b. I_T = 3.5 A
    c. I_{R_1} = 1.5 A
    d. I_{R_2} = 2 A
    e. P_{R_1} = 180 W
    f. P_{R_2} = 240 W
    g. P_T = 420 W
27. a. R_A = 120 Ω
    b. R_B = 160 Ω
    c. R_C = 200 Ω
    d. R_T = 51.1 Ω
    e. I_A = 0.417 A
    f. I_B = 0.313 A
    g. I_C = 0.25 A
    h. I_T = 0.978 A
    i. P_T = 48.9 W

28. a. $R_T = 300\ \Omega$
    b. $R_C = 1.4\ k\Omega$
    c. $I_A = 25$ mA
    d. $I_C = 21.4$ mA
    e. $I_B = 53.6$ mA
    f. $R_B = 560\ \Omega$
    g. $P_T = 3$ W

29. a. $V = 100$ V
    b. $I_{R2} = 1.33$ A
    c. $R_3 = 33.3\ \Omega$
    d. $I_T = 5.33$ A
    e. $P_{R1} = 100$ W
    f. $P_{R2} = 133$ W
    g. $P_{R3} = 300$ W
    h. $P_T = 533$ W

30. a. $I_1 = 0.8$
    b. $I_2 = 0.533$ A
    c. $I_3 = 0.4$ A
    d. $I_4 = 0.267$ A
    e. $I_5 = 0.667$ A
    f. $I_6 = 1.2$ A
    g. $I_7 = 2$ A
    h. $I_8 = 2$ A
    i. $I_9 = 1.2$ A
    j. $I_{10} = 0.667$ A
    k. $I_{11} = 0.267$ A

ANSWERS TO WORKBOOK, page 173

1. a. Formula: $\mathbf{V = I \times R}$
   Substitution: $\mathbf{V = 15\ A \times 20\ \Omega}$
   Solution: $\mathbf{V = 300\ V}$

   b. Formula: $\mathbf{P = I \times V}$
   Substitution: $\mathbf{P = 15\ A \times 300\ V}$
   Solution: $\mathbf{P = 4.5\ kW}$

2. a. Formula: $\mathbf{V = \frac{P}{I}}$
   Substitution: $\mathbf{V = \frac{15\ W}{5\ A}}$
   Solution: $\mathbf{V = 3\ V}$

   b. Formula: $\mathbf{R = \frac{V}{I}}$
   Substitution: $\mathbf{R = \frac{3\ V}{5\ A}}$
   Solution: $\mathbf{R = 0.6\ \Omega}$

3. a. Formula: $\mathbf{R = \frac{V}{I}}$
   Substitution: $\mathbf{R = \frac{220\ V}{0.06\ A}}$
   Solution: $\mathbf{R = 3.67\ k\Omega}$

   b. Formula: $\mathbf{P = I \times V}$
   Substitution: $\mathbf{P = 0.06\ A \times 220\ V}$
   Solution: $\mathbf{P = 13.2\ W}$

4. a. Formula: $\mathbf{I = \frac{V}{R}}$
   Substitution: $\mathbf{I = \frac{120\ V}{10\ \Omega}}$
   Solution: $\mathbf{I = 12\ A}$

   b. Formula: $\mathbf{P = I \times V}$
   Substitution: $\mathbf{P = 12\ A \times 120\ V}$
   Solution: $\mathbf{P = 1.44\ kW}$

5. a. Formula: $\mathbf{I = \frac{P}{V}}$
   Substitution: $\mathbf{I = \frac{0.3\ W}{50\ V}}$
   Solution: $\mathbf{I = 6\ mA}$

   b. Formula: $\mathbf{R = \frac{V}{I}}$
   Substitution: $\mathbf{R = \frac{50\ V}{0.006\ A}}$
   Solution: $\mathbf{R = 8.33\ k\Omega}$

6. a. Formula: $\mathbf{V = \sqrt{R \times P}}$
   Substitution: $\mathbf{V = \sqrt{5000\ \Omega \times 0.03\ W}}$
   Solution: $\mathbf{V = 12.2\ V}$

   b. Formula: $\mathbf{I = \frac{V}{R}}$
   Substitution: $\mathbf{I = \frac{12.2\ V}{5000\ \Omega}}$
   Solution: $\mathbf{I = 2.44\ mA}$

7. a. Formula: $\mathbf{V = I \times R}$
   Substitution: $\mathbf{V = 0.075\ A \times 20{,}000\ \Omega}$
   Solution: $\mathbf{V = 1.5\ kV}$

   b. Formula: $\mathbf{P = I \times V}$
   Substitution: $\mathbf{P = 0.075\ A \times 1500\ V}$
   Solution: $\mathbf{P = 113\ W}$

8. a. Formula: $\mathbf{V = \frac{P}{I}}$
   Substitution: $\mathbf{V = \frac{60\ mW}{3\ mA}}$
   Solution: $\mathbf{V = 20\ V}$

   b. Formula: $\mathbf{R = \frac{V}{I}}$
   Substitution: $\mathbf{R = \frac{20\ V}{0.003\ A}}$
   Solution: $\mathbf{R = 6.67\ k\Omega}$

9. a. Formula: $\mathbf{R = \frac{V}{I}}$
   Substitution: $\mathbf{R = \frac{100\ V}{0.01\ A}}$
   Solution: $\mathbf{R = 10\ k\Omega}$

   b. Formula: $\mathbf{P = I \times V}$
   Substitution: $\mathbf{P = 0.01\ A \times 100\ V}$
   Solution: $\mathbf{P = 1\ W}$

10. a. Formula: $I = \frac{V}{R}$
Substitution: $I = \frac{15\ V}{7500\ \Omega}$
Solution: $I = 2\ mA$
b. Formula: $P = I \times V$
Substitution: $P = 0.002\ A \times 15\ V$
Solution: $P = 30\ mW$

11. a. Formula: $I = \frac{V}{R}$
Substitution: $I = \frac{0\ V}{500\ \Omega}$
Solution: $I = 0\ A$
b. Formula: $P = I \times V$
Substitution: $P = 0\ A \times 0\ V$
Solution: $P = 0\ W$

12. a. Formula: $I = \frac{P}{V}$
Substitution: $I = \frac{8\ W}{20\ V}$
Solution: $I = 0.4\ A$
b. Formula: $R = \frac{V}{I}$
Substitution: $R = \frac{20\ V}{0.4\ A}$
Solution: $R = 50\ \Omega$

13. a. Formula: $V = I \times R$
Substitution: $V = 0.000015\ A \times 15{,}000\ \Omega$
Solution: $V = 0.225\ V$
b. Formula: $P = I \times V$
Substitution: $P = 0.000015\ A \times 0.225\ V$
Solution: $P = 3.38\ \mu W$

14. a. Formula: $V = \frac{P}{I}$
Substitution: $V = \frac{4\ W}{0.4\ A}$
Solution: $V = 10\ V$
b. Formula: $R = \frac{V}{I}$
Substitution: $R = \frac{10\ V}{0.4\ A}$
Solution: $R = 25\ \Omega$

15. a. Formula: $R = \frac{V}{I}$
Substitution: $R = \frac{35\ V}{5\ A}$
Solution: $R = 7\ \Omega$
b. Formula: $P = I \times V$
Substitution: $P = 5\ A \times 35\ V$
Solution: $P = 175\ W$

16. a. Formula: $I = \frac{V}{R}$
Substitution: $I = \frac{40\ V}{24\ \Omega}$
Solution: $I = 1.67\ A$
b. Formula: $P = I \times V$
Substitution: $P = 1.67\ A \times 40\ V$
Solution: $P = 66.8\ W$

17. a. Formula: $I = \frac{P}{V}$
Substitution: $I = \frac{9\ W}{9\ V}$
Solution: $I = 1\ A$
b. Formula: $R = \frac{V}{I}$
Substitution: $R = \frac{9\ V}{1\ A}$
Solution: $R = 9\ \Omega$

18. a. Formula: $V = \sqrt{R \times P}$
Substitution: $V = 2000\ \Omega \times 0.4\ W$
Solution: $V = 28.3\ V$
b. Formula: $I = \frac{V}{R}$
Substitution: $I = \frac{28.3\ V}{2000\ \Omega}$
Solution: $I = 14.2\ mA$

19. a. Formula: $V = I \times R$
Substitution: $V = 7\ A \times 250\ \Omega$
Solution: $V = 1.75\ kV$
b. Formula: $P = I \times V$
Substitution: $P = 7\ A \times 1750\ V$
Solution: $P = 12.3\ kW$

20. a. Formula: $R = \frac{V}{I}$
Substitution: $R = \frac{12\ V}{0.12\ A}$
Solution: $R = 100\ \Omega$
b. Formula: $P = I \times V$
Substitution: $P = 0.12\ A \times 12\ V$
Solution: $P = 1.44\ W$

21. a. Formula: $R_T = R_1 + R_2 + R_3$
Substitution: $R_T = 550\ \Omega + 350\ \Omega + 200\ \Omega$
Solution: $R_T = 1.1\ k\Omega$
b. Formula: $I_T = \frac{V_T}{R_T}$
Substitution: $I_T = \frac{200\ V}{1100\ \Omega}$
Solution: $I_T = 182\ mA$
c. Formula: $V_{R_L} = I_T \times R_L$
Substitution: $V_{R_1} = 0.182\ A \times 550\ \Omega$
Solution: $V_{R_1} = 100\ V$
d. Formula: $V_{R_L} = I_T \times R_L$
Substitution: $V_{R_2} = 0.182\ A \times 350\ \Omega$
Solution: $V_{R_2} = 63.7\ V$
e. Formula: $V_{R_L} = I_T \times R_L$
Substitution: $V_{R_3} = 0.182\ A \times 200\ \Omega$
Solution: $V_{R_3} = 36.4\ V$

f. Formula: $P_{R_1} = I_T^2 \times R_L$
Substitution: $P_{R_1} = (0.182\ A)^2 \times 550\ \Omega$
Solution: $P_{R_1} = 18.2\ W$

g. Formula: $P_{R_2} = I_T^2 \times R_L$
Substitution: $P_{R_2} = (0.182\ A)^2 \times 350\ \Omega$
Solution: $P_{R_2} = 11.6\ W$

h. Formula: $P_{R_3} = I_T^2 \times R_L$
Substitution: $P_{R_3} = (0.182\ A)^2 \times 200\ \Omega$
Solution: $P_{R_3} = 6.62\ W$

i. Formula: $P_T = P_{R_1} + P_{R_2} + P_{R_3}$
Substitution: $P_T = 18.2\ W + 11.6\ W + 6.62\ W$
Solution: $P_T = 36.4\ W$

22. a. Formula: $R_T = R_1 + R_2 + R_3$
Substitution: $R_T = 15\ \Omega + 30\ \Omega + 25\ \Omega$
Solution: $R_T = 70\ \Omega$

b. Formula: $I_T = \frac{V_{R_L}}{R_L}$
Substitution: $I_T = \frac{15\ V}{30\ \Omega}$
Solution: $I_T = 0.5\ A$

c. Formula: $V_{R_L} = I_T \times R_L$
Substitution: $V_{R_1} = 0.5\ A \times 15\ \Omega$
Solution: $V_{R_1} = 7.5\ V$

d. Formula: $V_{R_L} = I_T \times R_L$
Substitution: $V_{R_3} = 0.5\ A \times 25\ \Omega$
Solution: $V_{R_3} = 12.5\ V$

e. Formula: $V_T = V_{R_1} + V_{R_2} + V_{R_3}$
Substitution: $V_T = 7.5\ V + 15\ V + 12.5\ V$
Soluton: $V_T = 35\ V$

f. Formula: $P_{R_1} = I_T^2 \times R_L$
Substitution: $P_{R_1} = (0.5\ A)^2 \times 15\ \Omega$
Solution: $P_{R_1} = 3.75\ W$

g. Formula: $P_{R_2} = I_T^2 \times R_L$
Substitution: $P_{R_2} = (0.5\ A)^2 \times 30\ \Omega$
Solution: $P_{R_2} = 7.5\ W$

h. Formula: $P_{R_3} = I_T^2 \times R_L$
Substitution: $P_{R_3} = (0.5\ A)^2 \times 25\ \Omega$
Solution: $P_{R_3} = 6.25\ W$

i. Formula: $P_T = P_{R_1} + P_{R_2} + P_{R_3}$
Substitution: $P_T = 3.75\ W + 7.5\ W + 6.25\ W$
Solution: $P_T = 17.5\ W$

23. a. Formula: $R_T = \frac{V_T}{I_T}$
Substitution: $R_T = \frac{120\ V}{2\ A}$
Solution: $R_T = 60\ \Omega$

b. Formula: $R_3 = R_T - (R_1 + R_2 + R_4 + R_5 + R_6)$
Substitution: $R_3 = 60\Omega - (5\Omega + 5\Omega + 10\Omega + 15\Omega + 10\Omega)$
Solution: $R_3 = 15\ \Omega$

c. Formula: $P_T = I_T \times V_T$
Substitution: $P_T = 2\ A \times 120\ V$
Solution: $P_T = 240\ W$

24. a. Formula: $R_L = \frac{V_{R_L}}{I_T}$
Substitution: $R_1 = \frac{15\ V}{0.005\ A}$
Solution: $R_1 = 3\ k\Omega$

b. Formula: $R_T = R_1 + R_2 + R_3$
Substitution: $R_T = 3\ k\Omega + 2\ k\Omega + 5\ k\Omega$
Solution: $R_T = 10\ k\Omega$

c. Formula: $V_T = I_T \times R_T$
Substitution: $V_T = 0.005\ A \times 10{,}000\ \Omega$
Solution: $V_T = 50\ V$

d. Formula: $V_{R_L} = I_T \times R_L$
Substitution: $V_{R_2} = 0.005\ A \times 2000\ \Omega$
Solution: $V_{R_2} = 10\ V$

e. Formula: $V_{R_L} = I_T \times R_L$
Substitution: $V_{R_3} = 0.005\ A \times 5000\ \Omega$
Solution: $V_{R_3} = 25\ V$

f. Formula: $P_T = I_T \times V_T$
Substitution: $P_T = 0.005\ A \times 50\ V$
Solution: $P_T = 250\ mW$

25. a. Formula: $I_T = \sqrt{\frac{P_{R_L}}{R_L}}$
Substitution: $I_T = \sqrt{\frac{3\ W}{100\ \Omega}}$
Solution: $I_T = 173\ mA$

b. Formula: $R_T = \frac{V_T}{I_T}$
Substitution: $R_T = \frac{40\ V}{0.173\ A}$
Solution: $R_T = 231\ \Omega$

c. Formula: $R_2 = R_T - (R_1 + R_3)$
Substitution: $R_2 = 231\ \Omega - (80\ \Omega + 100\ \Omega)$
Solution: $R_2 = 51\ \Omega$

d. Formula: $P_{R_L} = I_T^2 \times R_L$
Substitution: $P_{R_1} = (0.173\ A)^2 \times 80\ \Omega$
Solution: $P_{R_1} = 2.39\ W$

e. Formula: $P_{R_L} = I_T^2 \times R_L$
Substitution: $P_{R_2} = (0.173\ A)^2 \times 51\ \Omega$
Solution: $P_{R_2} = 1.53\ W$

f. Formula: $P_T = I_T \times V_T$
Substitution: $P_T = 0.173\ A \times 40\ V$
Solution: $P_T = 6.92\ W$

26. a. Formula: $\frac{1}{R_T} = \frac{1}{R_1} + \frac{1}{R_2}$
Substitution: $\frac{1}{R_T} = \frac{1}{240\ \Omega} + \frac{1}{180\ \Omega}$
Solution: $R_T = 103\ \Omega$

b. Formula: $I_T = \frac{V_T}{R_T}$
Substitution: $I_T = \frac{50\ V}{103\ \Omega}$
Solution: $I_T = 0.485\ A$

c. Formula: $I_{R_L} = \frac{V_T}{R_L}$
Substitution: $I_{R_1} = \frac{50\ V}{240\ \Omega}$
Solution: $I_{R_1} = 0.208\ A$

d. Formula: $\mathbf{I_{R_L} = \dfrac{V_T}{R_L}}$

Substitution: $\mathbf{I_{R2} = \dfrac{50\ V}{180\ \Omega}}$

Solution: $\mathbf{I_{R2} = 0.278\ A}$

e. Formula: $\mathbf{P_{R_L} = I_{R_L} \times V_{R_L}}$
Substitution: $\mathbf{P_{R1} = 0.208\ A \times 50\ V}$
Solution: $\mathbf{P_{R1} = 10.4\ W}$

f. Formula: $\mathbf{P_{R_L} = I_{R_L} \times V_{R_L}}$
Substitution: $\mathbf{P_{R1} = 0.278\ A \times 50\ V}$
Solution: $\mathbf{P_{R1} = 13.9\ W}$

g. Formula: $\mathbf{P_T = I_T \times V_T}$
Substitution: $\mathbf{P_T = 0.485\ A \times 50\ V}$
Solution: $\mathbf{P_T = 24.3\ W}$

27. a. Formula: $\mathbf{R_A = R_1 + R_2 + R_3}$
Substitution: $\mathbf{R_A = 100\ \Omega + 150\ \Omega + 200\ \Omega}$
Solution: $\mathbf{R_A = 450\ \Omega}$

b. Formula: $\mathbf{R_B = R_4 + R_5}$
Substitution: $\mathbf{R_B = 75\ \Omega + 75\ \Omega}$
Solution: $\mathbf{R_B = 150\ \Omega}$

c. Formula: $\mathbf{R_C = R_6 + R_7 + R_8}$
Substitution: $\mathbf{R_C = 100\ \Omega + 75\ \Omega + 125\ \Omega}$
Solution: $\mathbf{R_C = 300\ \Omega}$

d. Formula: $\mathbf{\dfrac{1}{R_T} = \dfrac{1}{R_A} + \dfrac{1}{R_B} + \dfrac{1}{R_C}}$

Substitution: $\mathbf{\dfrac{1}{R_T} = \dfrac{1}{450\ \Omega} + \dfrac{1}{150\ \Omega} + \dfrac{1}{300\ \Omega}}$

Solution: $\mathbf{R_T = 81.8\ \Omega}$

e. Formula: $\mathbf{I_{R_L} = \dfrac{V_T}{R_L}}$

Substitution: $\mathbf{I_A = \dfrac{150\ V}{450\ \Omega}}$

Solution: $\mathbf{I_A = 0.333\ A}$

f. Formula: $\mathbf{I_{R_L} = \dfrac{V_T}{R_L}}$

Substitution: $\mathbf{I_B = \dfrac{150\ V}{150\ \Omega}}$

Solution: $\mathbf{I_B = 1\ A}$

g. Formula: $\mathbf{I_{R_L} = \dfrac{V_T}{R_L}}$

Substitution: $\mathbf{I_C = \dfrac{150\ V}{300\ \Omega}}$

Solution: $\mathbf{I_C = 0.5\ A}$

h. Formula: $\mathbf{I_T = \dfrac{V_T}{R_T}}$

Substitution: $\mathbf{I_T = \dfrac{150\ V}{81.8\ \Omega}}$

Solution: $\mathbf{I_T = 1.83\ A}$

i. Formula: $\mathbf{P_T = I_T \times V_T}$
Substitution: $\mathbf{P_T = 1.83\ A \times 150\ V}$
Solution: $\mathbf{P_T = 275\ W}$

28. a. Formula: $\mathbf{R_T = \dfrac{V_T}{I_T}}$

Substitution: $\mathbf{R_T = \dfrac{60\ V}{0.7\ A}}$

Solution: $\mathbf{R_T = 85.7\ \Omega}$

b. Formula: $\mathbf{R_B = R_2 + R_3}$
Substitution: $\mathbf{R_B = 150\ \Omega + 130\ \Omega}$
Solution: $\mathbf{R_B = 280\ \Omega}$

c. Formula: $\mathbf{I_{R_L} = \dfrac{V_T}{R_L}}$

Substitution: $\mathbf{I_B = \dfrac{60\ V}{280\ \Omega}}$

Solution: $\mathbf{I_B = 0.214\ A}$

d. Formula: $\mathbf{I_{R_L} = \dfrac{V_T}{R_L}}$

Substitution: $\mathbf{I_C = \dfrac{60\ V}{260\ \Omega}}$

Solution: $\mathbf{I_C = 0.231\ A}$

e. Formula: $\mathbf{I_A = I_T - (I_B + I_C)}$
Substitution: $\mathbf{I_A = 0.7\ A - (0.214\ A + 0.231\ A)}$
Solution: $\mathbf{I_A = 0.255\ A}$

f. Formula: $\mathbf{R_L = \dfrac{V_T}{I_{R_L}}}$

Substitution: $\mathbf{R_A = \dfrac{60\ V}{0.255\ A}}$

Solution: $\mathbf{R_A = 235\ \Omega}$

g. Formula: $\mathbf{P_T = I_T \times V_T}$
Substitution: $\mathbf{P_T = 0.7\ A \times 60\ V}$
Solution: $\mathbf{P_T = 42\ W}$

29. a. Formula: $\mathbf{V_T = I_{R_L} \times R_L}$
Substitution: $\mathbf{V_T = 0.3\ A \times 220\ \Omega}$
Solution: $\mathbf{V_T = 66\ V}$

b. Formula: $\mathbf{I_{R_L} = \dfrac{V_T}{R_L}}$

Substitution: $\mathbf{I_{R2} = \dfrac{66\ V}{660\ \Omega}}$

Solution: $\mathbf{I_{R2} = 0.1\ A}$

c. Formula: $\mathbf{R_L = \dfrac{V_T}{I_{R_L}}}$

Substitution: $\mathbf{R_3 = \dfrac{66\ V}{0.2\ A}}$

Solution: $\mathbf{R_3 = 330\ \Omega}$

d. Formula: $\mathbf{I_T = I_{R1} + I_{R2} + I_{R3}}$
Substitution: $\mathbf{I_T = 0.3\ A + 0.1\ A + 0.2\ A}$
Solution: $\mathbf{I_T = 0.6\ A}$

e. Formula: $\mathbf{P_{R_L} = I_{R_L} \times V_{R_L}}$
Substitution: $\mathbf{P_{R1} = 0.3\ A \times 66\ V}$
Solution: $\mathbf{P_{R1} = 19.8\ W}$

f. Formula: $\mathbf{P_{R_L} = I_{R_L} \times V_{R_L}}$
Substitution: $\mathbf{P_{R2} = 0.1\ A \times 66\ V}$
Solution: $\mathbf{P_{R2} = 6.6\ W}$

g. Formula: $P_{R_L} = I_{R_L} \times V_{R_L}$
Substitution: $P_{R_3} = 0.2\text{ A} \times 66\text{ V}$
Solution: $P_{R_3} = 13.2\text{ W}$

h. Formula: $P_T = I_T \times V_T$
Substitution: $P_T = 0.6\text{ A} \times 66\text{ V}$
Solution: $P_T = 39.6\text{ W}$

i. Formula: $R_T = \frac{V_T}{I_T}$
Substitution: $R_T = \frac{66\text{ V}}{0.6\text{ A}}$
Solution: $R_T = 110\ \Omega$

30. a. Formula: $I_{R_L} = \frac{V_T}{R_L}$
Substitution: $I_1 = \frac{300\text{ V}}{2000\ \Omega}$
Solution: $I_1 = 0.15\text{ A}$

b. Formula: $I_{R_L} = \frac{V_T}{R_L}$
Substitution: $I_2 = \frac{300\text{ V}}{3000\ \Omega}$
Solution: $I_2 = 0.1\text{ A}$

c. Formula: $I_{R_L} = \frac{V_T}{R_L}$
Substitution: $I_3 = \frac{300\text{ V}}{1000\ \Omega}$
Solution: $I_3 = 0.3\text{ A}$

d. Formula: $I_{R_L} = \frac{V_T}{R_L}$
Substitution: $I_4 = \frac{300\text{ V}}{4000\ \Omega}$
Solution: $I_4 = 75\text{ mA}$

e. Formula: $I_5 = I_1 + I_6$
Substitution: $I_5 = 0.15\text{ A} + 0.475\text{ A}$
Solution: $I_5 = 0.625\text{ A}$

f. Formula: $I_6 = I_2 + I_7$
Substitution: $I_6 = 0.1\text{ A} + 0.375\text{ A}$
Solution: $I_6 = 0.475\text{ A}$

g. Formula: $I_7 = I_3 + I_4$
Substitution: $I_7 = 0.3\text{ A} + 0.075\text{ A}$
Solution: $I_7 = 0.375\text{ A}$

h. Formula: $I_8 = I_1 + I_9$
Substitution: $I_8 = 0.15\text{ A} + 0.475\text{ A}$
Solution: $I_8 = 0.625$

i. Formula: $I_9 = I_2 + I_{10}$
Substitution: $I_9 = 0.1\text{ A} + 0.375\text{ A}$
Solution: $I_9 = 0.475\text{ A}$

j. Formula: $I_{10} = I_3 + I_4$
Substitution: $I_{10} = 0.3\text{ A} + 0.075\text{ A}$
Solution: $I_{10} = 0.375\text{ A}$

ANSWERS TO TEST MASTERS, page 171

1. 37.5 Ω
2. 150 V
3. 0.12 A
4. 4 A
5. 240 Ω
6. 2.4 Ω
7. 50 V
8. 2 A
9. 1800 W
10. 750 W
11. a. $R_T = 1000\ \Omega$
    b. $I_T = 0.5\text{ A}$
    c. $P_T = 250\text{ W}$
12. a. $R_T = 400\ \Omega$
    b. $I_T = 0.25\text{ A}$
    c. $V = 100\text{ V}$
13. a. $R_1 = 3\text{ k}\Omega$
    b. $R_T = 10\text{ k}\Omega$
    c. $V_{R_2} = 20\text{ V}$
    d. $P_{R_3} = 0.5\text{ W}$
14. a. $R_T = 300\ \Omega$
    b. $I_T = 0.2\text{ A}$
    c. $I_A = 67\text{ mA}$
    d. $I_B = 67\text{ mA}$
15. a. $R_2 = 40\ \Omega$
    b. $R_T = 18.5\ \Omega$
    c. $I_{R_1} = 0.15\text{ A}$
    d. $V_T = 12\text{ V}$
16. a. $I_1 = 1\text{ mA}$
    b. $I_2 = 0.4\text{ mA}$
    c. $I_3 = 2\text{ mA}$
    d. $I_4 = 3.4\text{ mA}$
    e. $I_5 = 2.4\text{ mA}$

Chapter 18
INTRODUCTION TO ESTIMATING

Objectives

After studying this chapter, the student will be able to:

- Calculate board feet given board length in inches or feet.
- Estimate wood needed for foundation sills.
- Estimate wood needed for wall plates.
- Estimate wood needed for door and window trim.
- Estimate wood needed for framing members.
- Estimate material requirements for sheet products.
- Estimate concrete material requirements.

ANSWERS TO TEXT, page 294

Board Measure

1. a. 6.7 bd. ft.
   b. 10.7 bd. ft.
   c. 3 bd. ft.
   d. 12 bd. ft.
   e. 5 bd. ft.
   f. 10 bd. ft.
   g. 12 bd. ft.
   h. 4 bd. ft.
   i. 9.7 bd. ft.
   j. 426.7 bd. ft.

2. a. 630 in.
   b. 120 ft.
   c. 432 in.
   d. 348 ft.
   e. 2160 in.
   f. 30 ft.
   g. 960 in.
   h. 390 ft.
   i. 2160 in.
   j. 12 ft.

Linear Measure

3. a. 124′2″
   Horizontal: 25′ + 10′ + 35′ = 70′
   Vertical: [28′ − (5 1/2″ × 2)] + [13′ − (5 1/2″ × 2)] + [(15′ + 5 1/2″) − 5 1/2″] = 27′1″ + 12′1″ + 15′ = 54′2″
   Total: 70′ + 54′2″ = 124′2″
   b. 4-10′, 3-14′, 3-16′
   Starting on top, going clockwise:
   10′ + 16′, 6′, 10′, 14′, 2-10′ + 16′, 2-14′ = 4-10′, 3-14′, 3-16′
4. 168′
   3 × 56′ = 168′
5. 22′4″
   4′7″ + 6′7″ + 4′7″ + 6′7″ = 20′28″ = 22′4″
6. 21′2″
   Top piece: 6′ + (5 1/2″ × 2) = 6′11″
   Side pieces: (6′8″ + 5 1/2″) × 2 = 14′3″
   Total: 6′11″ + 14′3″ = 21′2″

Framing Member Count

7. a. 19
   b. 31
   c. 16
   d. 9
   e. 8
   f. 9
   g. 20
   h. 5
   i. 12
   j. 9

Square Measure

8. 27 sheets

$$\frac{24 \text{ ft.} \times 36 \text{ ft.}}{32 \text{ ft.}^2/\text{sheet}} = 27 \text{ sheets}$$

9. 24 tiles

$$\frac{12 \text{ ft.} \times 16 \text{ ft.}}{8 \text{ ft.}^2/\text{tile}} = 24 \text{ tiles}$$

10. 24 sheets
   14-ft. walls: 14 ft. × 8 ft. × 2 = 224 ft.^{2}
   18-ft. walls: 18 ft. × 8 ft. × 2 = 288 ft.^{2}
   Ceiling: 14 ft. × 18 ft. = 252 ft.^{2}

$$\frac{224 \text{ ft.}^2 + 288 \text{ ft.}^2 + 252 \text{ ft.}^2}{32 \text{ ft.}^2/\text{sheet}} = 24 \text{ sheets}$$

11. a. 24 sheets

$$\text{Floor: } \frac{24 \text{ ft.} \times 32 \text{ ft.}}{32 \text{ ft.}^2/\text{sheet}} = 24 \text{ sheets}$$

   b. 14 sheets
   Front and rear:
   Rectangle: 32 ft. × 8 ft. × 2 = 512 ft.^{2}
   Openings: 2[(2 ft. × 2 ft.) + (3 ft. × 7 ft.) + (3 ft. × 6 ft.)] = 86 ft.^{2}

$$\frac{512 \text{ ft.}^2 - 86 \text{ ft.}^2}{32 \text{ ft.}^2/\text{sheet}} = 13.3 \text{ sheets} \Rightarrow 14 \text{ sheets}$$

   c. 15 sheets
   Left and right sides:
   Rectangle: 24 ft. × 8 ft. × 2 = 384 ft.^{2}
   Triangle: 1/2 × 6 ft. × 24 ft. × 2 = 144 ft.^{2}
   Openings: 4 ft. × 4 ft. × 4 = 64 ft.^{2}

$$\frac{(384 \text{ ft.}^2 + 144 \text{ ft.}^2) - 64 \text{ ft.}^2}{32 \text{ ft.}^2/\text{sheet}}$$

$$= 14.5 \text{ sheets} \Rightarrow 15 \text{ sheets}$$

   d. 27 sheets
   Roof:
   Length of slope:

$$c = \sqrt{(6 \text{ ft.})^2 + (12 \text{ ft.})^2} = 13.42 \text{ ft.}$$

$$\frac{13.42 \text{ ft.} \times 32 \text{ ft.} \times 2}{32 \text{ ft.}^2/\text{sheet}}$$

$$= 26.84 \text{ sheets} \Rightarrow 27 \text{ sheets}$$

Cubic Measure

12. 0.8 yd.^{3}

$$\frac{8 \text{ ft.} \times 8 \text{ ft.} \times 0.33 \text{ ft.}}{27} = 0.782 \text{ yd.}^3$$

13. a. 10 yd.^{3}

$$\frac{32 \text{ ft.} \times 20 \text{ ft.} \times 0.42 \text{ ft.}}{27} = 9.96 \text{ yd.}^3$$

   b. \$600

$$10 \cancel{\text{yd.}^3} \left(\frac{\$60}{\cancel{\text{yd.}^3}}\right) = \$600$$

Building Estimate

14. a. 7 yd.^{3}
   Concrete foundation:
   Total length: 2(24 ft.) + 2(23.5 ft.) = 95 ft.

$$\frac{95 \text{ ft.} \times 4 \text{ ft.} \times 0.5 \text{ ft.}}{27} = 7.04 \text{ yd.}^3$$

   b. 6.7 yd.^{3}
   Concrete floor:

$$\frac{23.5 \text{ ft.} \times 23.5 \text{ ft.} \times 0.33 \text{ ft.}}{27} = 6.7 \text{ yd.}^3$$

   c. 78 studs
   Wall studs:
   Perimeter less door: (24 ft. × 4) − 8 ft. = 88 ft.

$$(88 \cancel{\text{ft.}}) \left(\frac{12 \text{ in.}}{1 \cancel{\text{ft.}}}\right) = 1056 \text{ in.}$$

$$\frac{1056 \text{ in.}}{16 \text{ in.}} + 1 = 67 \text{ studs}$$

   Extra studs:
   4 corners: 2 studs × 4 = 8 studs
   Door opening: 3 studs
   Total: 67 + 8 + 3 = 78 studs

d. 280 ft.
Plates:
Walls: 24 ft. × 3 = 72 ft.
3 walls: 72 ft. × 3 = 216 ft.
Wall/door: (top) 24 ft. × 2 = 48 ft.
(bottom) 24 ft. − 8 ft. = 16 ft.
Total: 216 ft. + 48 ft. + 16 ft. = 280 ft.

e. 38 rafters
Rafters:

$$24\text{ ft.}\left(\frac{12\text{ in.}}{\text{ft.}}\right) = 288\text{ in.}$$

$$\frac{288\text{ in.}}{16\text{ in.}} + 1 = 19\text{ rafters}$$

2 sides: 19 rafters × 2 = 38 rafters

f. 23 sheets
Walls:
4 walls: 24 ft. × 8 ft. × 4 = 768 ft.^{2}
Door: 8 ft. × 7 ft. = 56 ft.^{2}

$$\frac{768\text{ ft.}^2 - 56\text{ ft.}^2}{32\text{ ft.}^2/\text{sheet}} = 22.3\text{ sheets} \Rightarrow 23\text{ sheets}$$

g. 18 sheets
Roof:

$$\frac{24\text{ ft.} \times 24\text{ ft.}}{32\text{ ft.}^2/\text{sheet}} = 18\text{ sheets}$$

ANSWERS TO WORKBOOK, page 191

Board Measure

1. a. 8.89 bd. ft.
   b. 24 bd. ft.
   c. 7.08 bd. ft.
   d. 28 bd. ft.
   e. 43.75 bd. ft.
   f. 16.67 bd. ft.
   g. 1.33 bd. ft.
   h. 4 bd. ft.
   i. 18.33 bd. ft.
   j. 16 bd. ft.
2. a. 648 in.
   b. 90 ft.
   c. 2412 in.
   d. 72 ft.
   e. 5280 in.
   f. 1710 ft.
   g. 18,600 in.
   h. 99 ft.
   i. 720 in.
   j. 144 ft.

Linear Measure

3. a. 278′2″
   Horizontal: 80′ + 60′ + 20′ = 160′
   Vertical: [60′ − (5 1/2″ × 2)] + [40′ − (5 1/2″ × 2)] + [20′ + 5 1/2″) − 5 1/2″] = 59′1″ + 39′1″ + 20′ = 118′2″
   Total: 160′ + 118′2″ = 278′2″
   b. 28-10′
   Starting on top, going clockwise:
   8-10′, 4-10′, 2-10′, 2-10′, 6-10′, 6-10′ = 28-10′
4. 120′
   3 × 40′ = 120′
5. 27′6″
   7′4 1/2″ + 12′9″ + 7′4 1/2″ = 26′18″ = 27′6″
6. 12′4″
   2′7″ + 3′7″ + 2′7″ + 3′7″ = 10′28″ = 12′4″

Framing Member Count

7. a. 10
   b. 46
   c. 21
   d. 11
   e. 17
   f. 12
   g. 21
   h. 5
   i. 17
   j. 10

Square Measure

8. 90 sheets

$$\frac{48\text{ ft.} \times 60\text{ ft.}}{32\text{ ft.}^2/\text{sheet}} = 90\text{ sheets}$$

9. 48 tiles

$$\frac{16\text{ ft.} \times 24\text{ ft.}}{8\text{ ft.}^2/\text{tile}} = 48\text{ tiles}$$

10. 22 sheets
    14-foot walls: 14 ft. × 8 ft. × 2 = 224 ft.^{2}
    16-foot walls: 16 ft. × 8 ft. × 2 = 256 ft.^{2}
    Ceiling: 14 ft. × 16 ft. = 224 ft.^{2}

$$\frac{224\text{ ft.}^2 + 256\text{ ft.}^2 + 224\text{ ft.}^2}{32\text{ ft.}^2/\text{sheet}} = 22\text{ sheets}$$

11. a. 18 sheets
    Floor:

$$\frac{24\text{ ft.} \times 24\text{ ft.}}{32\text{ ft.}^2/\text{sheet}} = 18\text{ sheets}$$

   b. 12 sheets
   South side:
   Rectangle: 24 ft. × 12 ft. = 288 ft.^{2}
   Triangle: 1/2 × 8 ft. × 24 = 96 ft.^{2}

$$\frac{288\text{ ft.}^2 + 96\text{ ft.}^2}{32\text{ ft.}^2/\text{sheet}} = 12\text{ sheets}$$

   c. 10 sheets
   North side:
   Area (same as south side): 384 ft.^{2}
   Openings: 8 ft. × 8 ft. = 64 ft.^{2}

$$\frac{384\text{ ft.}^2 - 64\text{ ft.}^2}{32\text{ ft.}^2/\text{sheet}} = 10\text{ sheets}$$

   d. 8 sheets
   East side:
   Rectangle: 24 ft. × 12 ft. = 288 ft.^{2}
   Openings: (3 ft. × 7 ft.) + (4 ft. × 4 ft.) = 37 ft.^{2}

$$\frac{288\text{ ft.}^2 - 37\text{ ft.}^2}{32\text{ ft.}^2/\text{sheet}} = 7.8\text{ sheets}$$

e. 8 sheets
West side:
Rectangle (same as east side): 288 ft.^{2}
Openings: 4 ft. × 4 ft. × 2 = 32 ft.^{2}

$$\frac{288\text{ ft.}^2 - 32\text{ ft.}^2}{32\text{ ft.}^2/\text{sheet}} = 8\text{ sheets}$$

f. 22 sheets
Roof:
Length of slope:

$$c = \sqrt{(8\text{ ft.})^2 + (12\text{ ft.})^2} = 14.4\text{ ft.}$$

$$\frac{14.4\text{ ft.} \times 24\text{ ft.} \times 2}{32\text{ ft.}^2/\text{sheet}} = 21.6\text{ sheets}$$

Cubic Measure

12. 0.1 yd.^{3}

$$\frac{3\text{ ft.} \times 5\text{ ft.} \times 0.25\text{ ft.}}{27} = 0.139\text{ yd.}^3$$

13. a. 17.6 cu. yd.

$$\frac{36\text{ ft.} \times 40\text{ ft.} \times 0.33\text{ ft.}}{27} = 17.6\text{ yd.}^3$$

b. \$968

$$17.6\ \cancel{\text{yd.}^3}\left(\frac{\$55}{\cancel{\text{yd.}^3}}\right) = \$968$$

Building Estimate

14. a. 0.8 yd.^{3}
Concrete floor:

$$\frac{10\text{ ft.} \times 12\text{ ft.} \times 0.17\text{ ft.}}{27} = 0.756\text{ yd.}^3$$

b. 45 studs
Wall studs:
Perimeter: (10 ft. × 2) + (12 ft. × 2) = 44 ft.

$$44\ \cancel{\text{ft.}}\left(\frac{12\text{ in.}}{1\ \cancel{\text{ft.}}}\right) = 528\text{ in.}$$

$$\left(\frac{528\text{ in.}}{16\text{ in.}}\right) + 1 = 34\text{ studs}$$

4 corners: 2 studs × 4 = 8 studs
Door opening: 3 studs
Total: 34 + 8 + 3 = 45 studs

c. 10 rafters
Rafters:

$$12\ \cancel{\text{ft.}}\left(\frac{12\text{ in.}}{1\ \cancel{\text{ft.}}}\right) = 144\text{ in.}$$

$$\frac{144\text{ in.}}{16\text{ in.}} + 1 = 10\text{ rafters}$$

d. 10 sheets
Walls:
10-ft. walls: 10 ft. × 8 ft. × 2 = 160 ft.^{2}
12-ft. walls: 12 ft. × 8 ft. × 2 = 192 ft.^{2}
Door: 6 ft. × 6 ft. = 36 ft.^{2}

$$\frac{(160\text{ ft.}^2 + 192\text{ ft.}^2) - 36\text{ ft.}^2}{32\text{ ft.}^2/\text{sheet}} = 9.9\text{ sheets}$$

e. 4 sheets
Roof:

$$\frac{10\text{ ft.} \times 12\text{ ft.}}{32\text{ ft.}^2/\text{sheet}} = 3.75\text{ sheets}$$

ANSWERS TO TEST MASTERS, page 179

1. a. 5.6 bd. ft.
b. 60 bd. ft.
c. 6.7 bd. ft.
d. 32.2 bd. ft.
e. 48 bd. ft.

2. a. 432 in.
b. 96 ft.
c. 2362 in.
d. 225 ft.
e. 36 ft.

3. a. 210′2″
60 ft. × 2 = 120 ft.
45′1″ × 2 = 90′2″
120′ + 90′2″ = 210′2″
b. 18 2 × 6s

$$210.17\ \cancel{\text{ft.}}\left(\frac{1\text{ board}}{12\ \cancel{\text{ft.}}}\right) = 17.5\text{ boards}$$

4. 138′
46 ft. × 3 = 138 ft.

5. 17′8″
3′11″ + 3′11″ + 4′11″ + 4″11′
= 14′44″ = 17′8″

6. a. 13
b. 17
c. 21
d. 13
e. 6

7. 36 sheets

$$\frac{32\text{ ft.} \times 36\text{ ft.}}{32\text{ ft.}^2/\text{sheet}} = 36\text{ sheets}$$

8. 40 tiles

$$\frac{16\text{ ft.} \times 20\text{ ft.}}{8\text{ ft.}^2/\text{tile}} = 40\text{ tiles}$$

9. 4.8 yd.^{3}

$$\frac{6\text{ ft.} \times 32\text{ ft.} \times 0.67\text{ ft.}}{27} = 4.76\text{ yd.}^3$$

10. 10.6 yd.^{3}

$$\frac{36\text{ ft.} \times 24\text{ ft.} \times 0.33\text{ ft.}}{27} = 10.56\text{ yd.}^3$$

Test Masters for use with

Applied Mathematics

Chapter 1

SOLVING WORD PROBLEMS

Test Score ______________

Name ______________________________

Date ______________ Class ______________

Complete each of the following problems. Show your work and place answers on the lines provided. Include units of measure (for example, inches, feet, minutes, hours, etc.) with the final answer.

1. The total power consumption on a circuit was measured to be 5680 watts. Devices of the circuit that consume a total of 3500 watts of power are changed to another circuit. What is total power consumption on the original circuit after the change?

1.______________________

2. Oak flooring is delivered to a building site. There are 164 boards in the shipment. A carpenter uses 95 boards and has an additional 256 delivered for the next day. Two carpenters use 310 more boards and finish the job. How many boards were left unused?

2.______________________

3. In an auto repair shop, 55 cars have an oil change in one week. If each car contains 4 quarts of oil, how many quarts of dirty oil must be disposed of at the end of the week?

3.______________________

4. A drafting job pays $432 for a 48 hour workweek. What is the rate of pay per hour? (Assume no overtime is paid.)

4.____________________

5. A nurse finds a patient weighs 195 pounds. After dieting for a few weeks, the patient returns and weighs 178 pounds. The nurse says 15 more pounds must be lost before stopping the diet. How much will the patient weigh at the end of the diet?

5.____________________

6. A certain stairway will have a rise of 8 inches per step. How many steps will be needed if the stairway height is 32 inches?

6.____________________

7. A robot can install 26 components in two hours. If three more robots are put into service, how many components could be installed in 8 hours?

7.____________________

8. A certain used car costs $8500. In addition, tax is $425, registration is $60, insurance payment is $135, and dealer fee is $50. With a down payment of $3000, what amount will cover the remaining expenses?

8.____________________

9. A certain house in rural New England used the following amounts of electricity (measured in kilowatt-hours): January, 744; February, 672; March, 682; April, 600; May, 528; June, 527; July, 434; August, 403; September, 450; October, 620; November, 630; and December, 682. What is the average monthly kilowatt-hours used?

9.____________________

10. An electric water pump can fill a 1000-gallon, cattle-watering tank in 40 minutes. If the cattle drink 50 gallons while the tank fills, what is the additional time required to fill the tank?

10.____________________

Chapter 2

FRACTIONS

Test Score ______________

Name ______________________________

Date ______________ Class ______________

Complete each of the following problems. Show your work and place answers on the lines provided.

1. Change to a mixed number:

   $\frac{35}{6} =$ ______________

2. Change to an improper fraction:

   $3\frac{4}{5} =$ ______________

3. Change to an equivalent fraction having the denominator on the right:

   $\frac{5}{6} = \frac{\quad}{12}$

4. Change to an equivalent fraction having the denominator on the right:

$$\frac{3}{7} = \frac{\quad}{28}$$

5. Reduce to lowest terms:

$$\frac{14}{36} =$$ ____________________

6. Reduce to lowest terms:

$$\frac{36}{16} =$$ ____________________

7. Arrange in descending order; give the LCD:

$$\frac{5}{8} \quad \frac{9}{16} \quad \frac{3}{4} \quad \frac{23}{32}$$

_____ _____ _____ _____ LCD _____

8. Arrange in descending order; give the LCD:

$\frac{2}{3}$ $\frac{5}{12}$ $\frac{1}{4}$ $\frac{1}{2}$ _____ _____ _____ _____ LCD _____

9. Add (give answer in lowest terms):

$\frac{2}{5} + \frac{4}{7} =$ ____________________

10. Subtract (give answer in lowest terms):

$\frac{3}{4} - \frac{3}{8} =$ ____________________

11. Multiply (give answer in lowest terms):

$\frac{1}{2} \times \frac{2}{3} =$ ____________________

12. Divide (give answer in lowest terms):

$\frac{5}{6} \div \frac{4}{9} =$ ____________________

13. Add (give answer in lowest terms):

$2\frac{3}{12} + 6\frac{11}{24} =$ ____________________

14. Subtract (give answer in lowest terms):

$3\frac{1}{3} - 1\frac{3}{4} =$ ____________________

15. Multiply (give answer in lowest terms):

$4\frac{2}{5} \times 2\frac{5}{6} =$ ____________________

16. Divide (give answer in lowest terms):

$5\frac{1}{2} \div 6\frac{3}{7} =$ ____________________

17. A wall of a garage is framed with 2 x 4s (actual width is 3 1/2 inches). Inside is 5/8-inch, fire-code dry wall. Outside is 1/2-inch plywood and 3/8-inch siding. What is the total thickness of the wall? 17.____________________

18. With the drawing of the garage shown, determine the width of the overhead door. 18.____________________

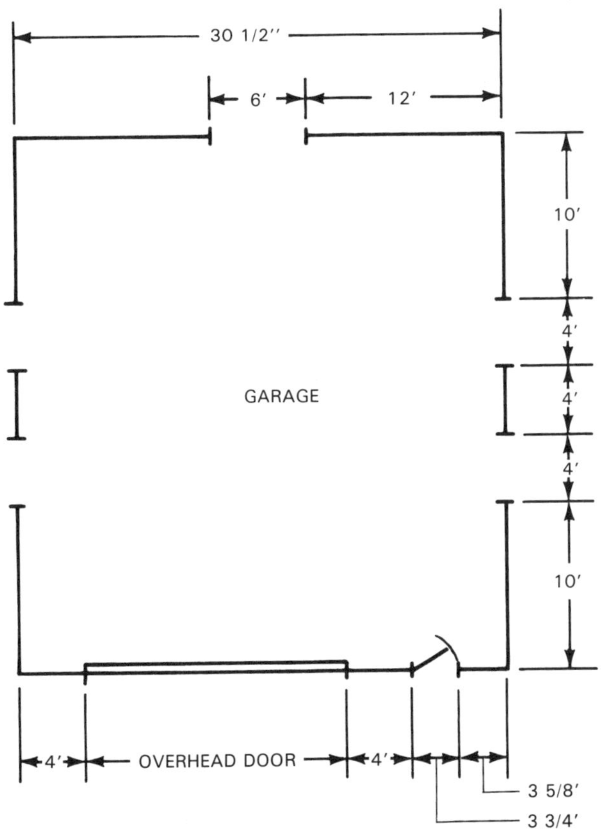

19. A black walnut plank will be smoothed in a planer. If the machine is set up to remove 3/16 of an inch, how much wood will be removed with 4 passes?

19. ____________________

20. An electrician used an average of 10 2/3 feet of wire per fixture in wiring the ceiling lights of a grocery store. At this rate:
    (a) How many fixtures could be connected with a 500-foot roll?
    (b) How much, if any, would be left over?

20a. ____________________

b. ____________________

Chapter 3

DECIMAL FRACTIONS

Test Score ______________

Name ______________________________

Date ______________ Class ______________

Complete each of the following problems. Show your work and place answers on the lines provided.

Write each digit times its place value.

	Whole Numbers	Decimal Numbers
1. 627.385 =	______ × ______	______ × ______
	______ × ______	______ × ______
	______ × ______	______ × ______

Write as a fraction, with the denominator being the place value of the least significant figure. Do *not* reduce.

2. 0.7 = ______________ 3. 0.38 = ______________

4. 0.255 = ______________ 5. 0.0061 = ______________

Round to 3 significant figures.

6. 275.3988 = ______________ 7. 95.89556 = ______________

8. 4.006029 = ______________ 9. 0.125099 = ______________

10. 0.003874 = ______________

Round to 3 decimal places.

11. 2.223956 = ______________ 12. 0.66666 = ______________

13. 0.333333 = ______________ 14. 0.43750 = ______________

15. 0.687499 = ______________ 16. 5.01167 = ______________

Perform the indicated operations.
Round division problems to 3 significant figures.

17. 56.09 + 1.3 = ______________ 18. 27.3 − 15.17 = ______________

19. $3.5 \times 2.04 =$ ____________________

20. $0.46 \div 0.8 =$ ____________________

21. $0.016 + 0.207 =$ ____________________

22. $0.386 - 0.0924 =$ ____________________

23. $0.025 \times 3.4 =$ ____________________

24. $1.6 - 0.077 =$ ____________________

25. $10 \div 0.26 =$ ____________________

26. $1.1 \div 33 =$ ____________________

Change the decimal numbers to fractions. Reduce to lowest terms. Change the fractions to decimal numbers Round to 3 decimal places (if necesary).

27. $2.36 =$ ____________________

28. $0.6 =$ ____________________

29. $\frac{9}{16}$ = ____________

30. $\frac{7}{8}$ = ____________

31. Using a micrometer, a machine part is found to measure 25 thousandths of an inch larger than it should. If the specified dimension is 685 thousandths of an inch, what is the actual dimension of the machine part? (Give answer in decimal form.)

31.____________

32. A paint-spraying machine can apply 0.37 of a gallon of paint per minute. How many gallons of paint will be used in 45 minutes?

32.____________

33. A 12.6 foot copper pipe is cut into four equal-length pieces. How long is each piece?

33.____________

Chapter 4

PERCENTAGE

Test Score ______________

Name ______________________________

Date ______________ Class ______________

Complete each of the following problems. Show your work and place answers on the lines provided.

	PERCENT	DECIMAL	FRACTION
1.	a. ______	b. ______	$\frac{3}{4}$
2.	a. ______	0.8	b. ______
3.	$66\frac{2}{3}\%$	a. ______	b. ______
4.	a. ______	b. ______	$\frac{3}{8}$
5.	4.5%	a. ______	b. ______

Solve the following percentage problems.

6. What number is 35% of 680? 6.______________

7. What percent of 36 is 18? 7.______________

8. Sixty-five (65) is 75% of what number?

8. ____________________

9. Find 40% of 8.

9. ____________________

10. What percent of 40 is 120?

10. ____________________

11. Ninety (90) is 200% of what number?

11. ____________________

12. With a sales tax on merchandise of 3.5%, how much tax is charged for a $15 item?

12.____________________

13. If 100 days of a 180 day school year have already gone by, what percentage of the school year remains?

13.____________________

14. In a small midwestern town, 1300 people voted in an election. If the remaining 700 registered voters did not participate, what was the percent of voter turnout?

14.____________________

15. A factory worker spends 1.5 hours of an 8-hour day on lunch and coffee break. What percentage of the workday is this?

15.____________________

Determine the maximum and minimum amount based on the percent tolerance. Do *not* round answers.

16. 490 feet ± 4%.

maximum = ________________

minimum = ________________

17. 5600 ohms ± 15%.

maximum = ________________

minimum = ________________

18. 0.35 liters ± 10%.

maximum = ________________

minimum = ________________

19. 0.0068 inches ± 5.5%.

maximum = ____________

minimum = ____________

20. 7.5 grams ± 0.3%.

maximum = ____________

minimum = ____________

Chapter 5

PERSONAL FINANCES

Test Score ______________

Name ______________________________

Date ______________ Class ______________

Complete each of the following problems. Show your work and place answers on the lines provided. Round answers to the nearest whole cent (two decimal places) where applicable.

1. How much change would be returned from a $20 bill on a purchase of 10 feet of chain costing $1.48 per foot? 1. ______________

2. Determine the gross income of a framing carpenter who earns $8.20 per hour for a 48-hour week. (Overtime is time and a half for hours over 40.) 2. ______________

3. A pipefitter, expecting a 10% raise, determines paycheck deductions to be 30%. This is figured from an old pay stub showing a gross pay of $580. What will be the expected net pay, after the raise, using the old pay stub as a basis for an estimate? 3. ______________

4. If sales tax is 7.5%, what will be the tax charged to purchase a television costing $385.00? 4.______

5. In a town with a property tax rate of 35 mils, what is the yearly tax on a garage valued at $12,500? 5.______

6. What is the simple interest earned in 9 months on $800 if the annual interest rate is 9%. 6.______

7. Determine the balance at the end of one year if $600 is deposited in a savings account earning 7% interest, compounded semiannually. 7.______

8. Estimate the monthly payment when \$4500 is borrowed for 48 months with an APR of 8%.

8.____________________

9. If the wholesale cost of an electric drill is \$25, what is the percent markup for a retail price of \$40?

9.____________________

10. A store advertises a 30% discount on shoes. If the retail price is \$35, what is the sale price?

10.____________________

TEST • TEST • TEST • TEST • TEST • TEST • TES

Chapter 6

GRAPHS

Test Score ______________

Name ______________________________

Date ______________ Class ______________

Complete each of the following problems. Show your work and place answers on the lines provided. Include units where necessary.

Use the bar graph shown to answer Questions 1 through 4.

A CAR DEALER'S SALES, BY MONTH

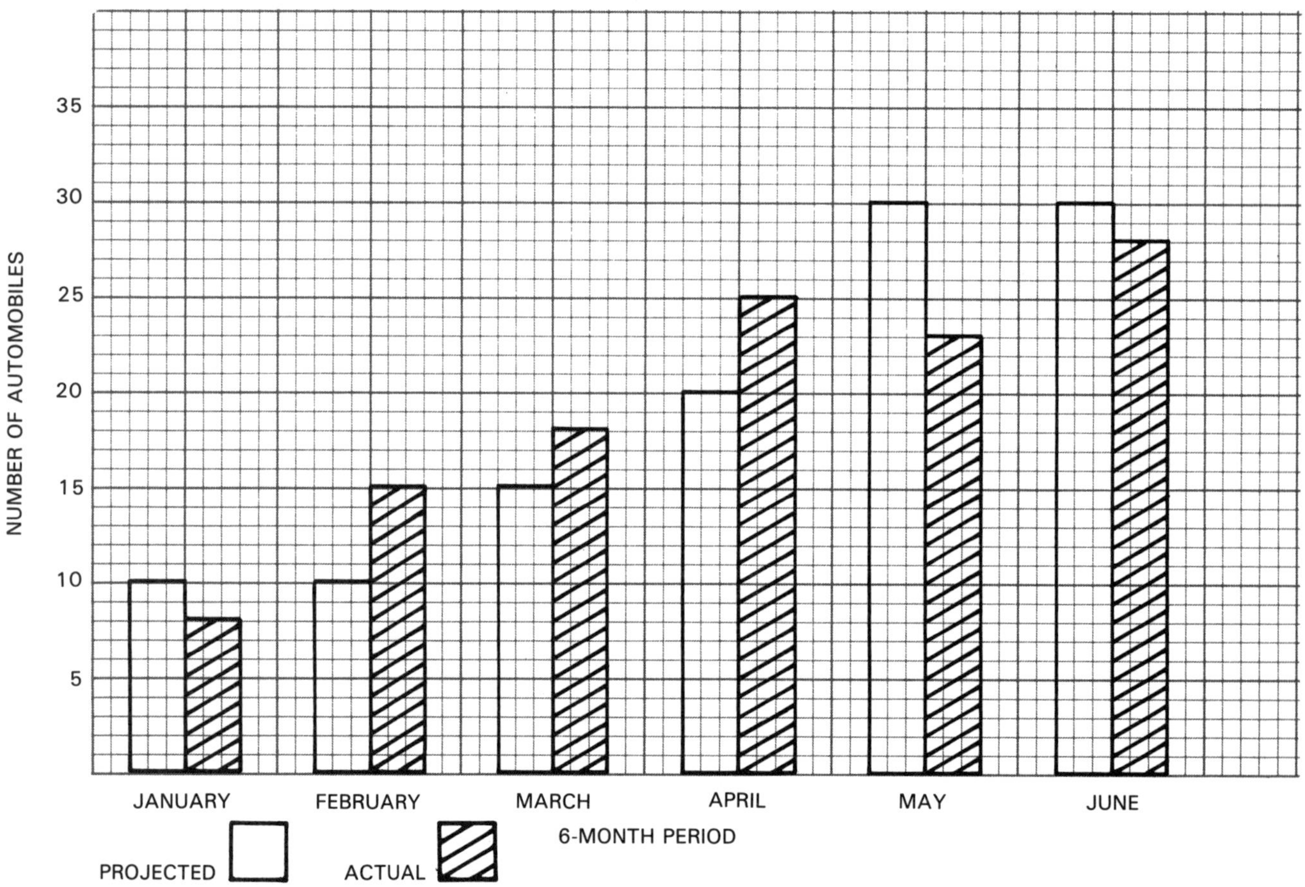

1. What was the total number of cars projected to be sold during the 6-month period? 1. ______________

2. What was the actual number of cars sold during this period? 2. ______________

3. During the month of May, the actual sales were under the projected by how many? 3. ______________

4. During the month of February, the actual sales were over the projected by how many? 4. ______________

5. Use the grid provided and the following information to plot a bar graph. 5.____________________

STUDENT POPULATION, BY GRADE LEVEL

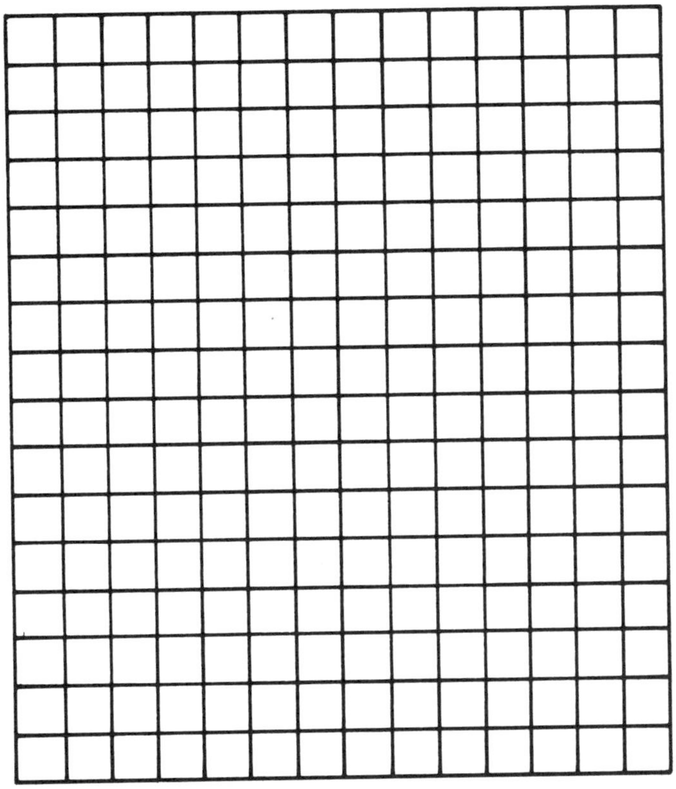

- Student population at the high school:

  9th grade ⇒ 300 students
  10th grade ⇒ 325 students
  11th grade ⇒ 270 students
  12th grade ⇒ 250 students

- Dependent variable is number of students—increment lines by 25.
- Independent variable is grade level.
- Label axes.

Use the line graph shown to answer Questions 6 through 9.

SINGLE FAMILY HOUSING SALES

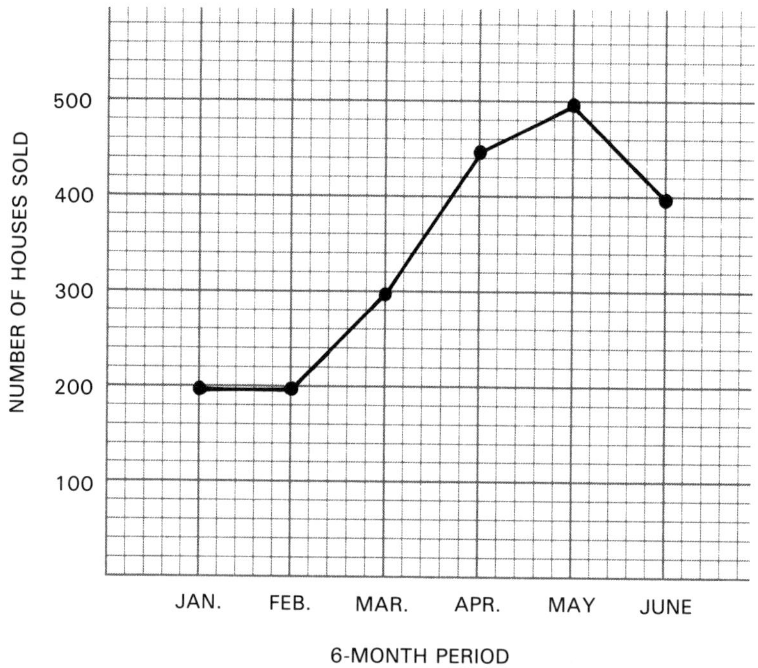

6. What is the trend of house sales during the spring months? 6. ____________

7. Which two months had equal sales? 7. ____________

8. What was the total number of houses sold? 8. ____________

9. If the average selling price of a house was $150,000, what was the dollar value of the housing market during this 6-month period? 9. ____________

10. Use the grid provided and the following information to plot a line graph.

AVERAGE MONTHLY RAINFALL

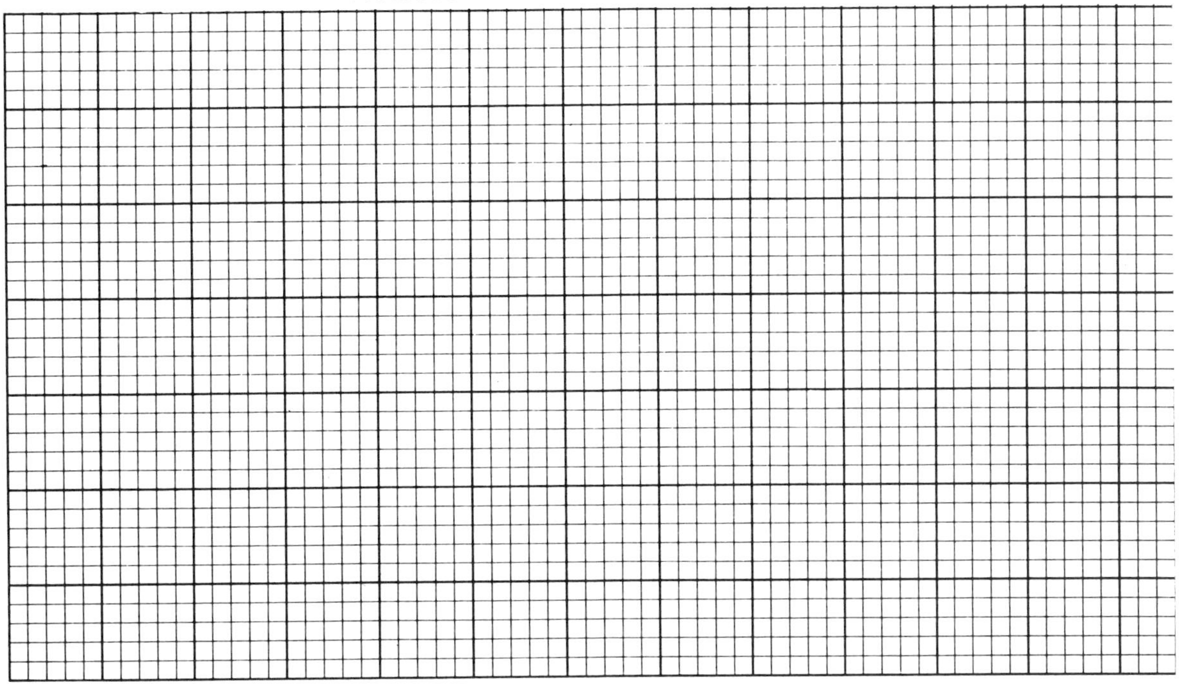

- Average monthly rainfall (inches):

January	= 6	July	= 5
February	= 3	August	= 2
March	= 10	September	= 20
April	= 18	October	= 25
May	= 22	November	= 13
June	= 15	December	= 8

- Dependent variable is rainfall—increment major divisions by 5.
- Independent variable is the month of the year.
- Label axes.

Use the circle graph shown to answer Questions 11 through 14.

CAUSES OF DEATH, AGES 15-25
(1,000,000 FATALITIES)

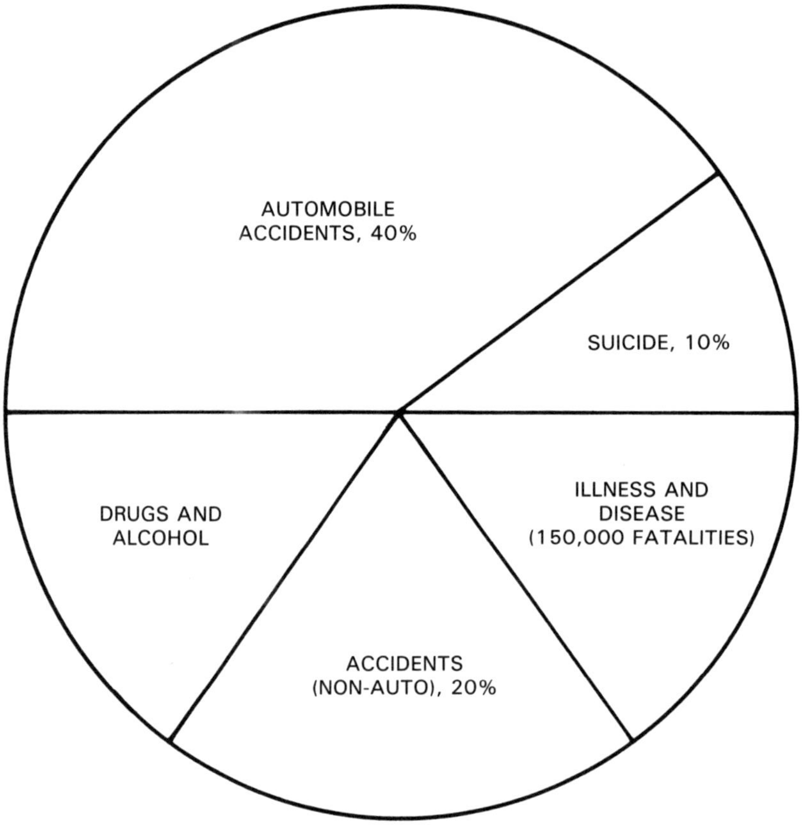

11. What is the percentage of fatalities caused by drugs and alcohol? 11. ____________

12. With the statistics given, how many people died from automobile accidents? 12. ____________

13. How many people died from suicide? 13. ____________

14. What is the percentage of fatalities caused by illness and disease? 14. ____________

15. Use the circle shown and the following information to create a circle graph. Use a protractor so that sectors are the correct size and label each sector. Use the space provided, for calculations.

PRODUCTION OF PAPER PRODUCTS

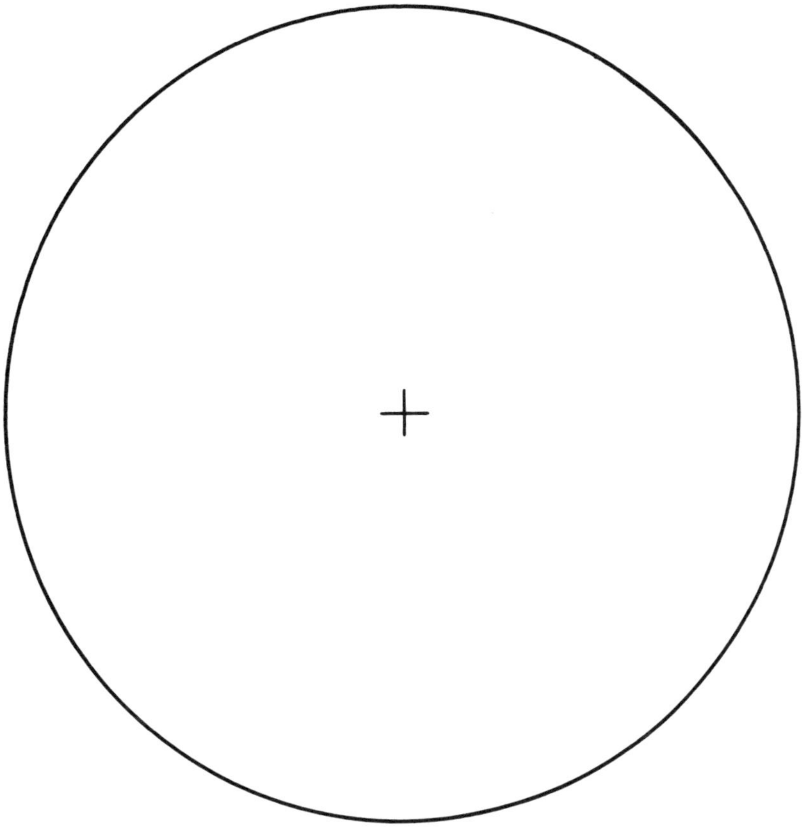

Production of paper products:

A = Boxes/packaging, 45%.

B = Writing paper, 30%.

C = Tissues, 5%.

D = Drawing paper, 15%.

E = Other, ?

Chapter 7

UNITS OF MEASURE

Test Score ______________

Name ______________________________

Date ______________ Class ______________

Complete each of the following problems. Show your work and place answers on the lines provided. Obtain conversion factors from Technical Section B or use Fig. 7-1 of the textbook.

Convert the measurement given to the new unit shown.

1. 88 inches ________________ yards

2. 356 square inches ________________ $ft.^2$

3. 128 ounces ________________ pounds

4. 3 quarts ________________ fluid ounces

5. 500 meters ________________ hectometers

6. 6 cubic meters ________________ dm^3

7. 2000 centigrams ________________ grams

8. 20 kiloliters ________________ milliliters

9. 2 weeks ________________ minutes

10. 96 hours ________________ days

11. 70 miles per hour ________________ ft./sec.

12. 30 feet per minute ________________ yards/min.

13. 120 liters per kilometer ________________ L/m

14. 480 grams per book ________________ kg/book

15. 32 feet per second ________________ miles/hour

16. 60 miles per hour ________________ km/hour

17. 150 pounds ________________ kg

18. 25 liters ________________ quarts

19. 15 inches ________________ cm

20. 45 square meters ______________ ft.^{2}

Perform the indicated arithmetic operations. Answers in compound denominate numbers should be reduced to the lowest terms.

21. 35 gallons
+ 105 quarts

21. ______________

22. 36 yards, 15 feet, 21 inches
+ 3 yards, 12 feet, 33 inches

22. ______________

23. 2 tons, 500 pounds
− 1 ton, 1000 pounds

23. ______________

24. (3 gallons, 4 quarts, 5 pints) × 2

24.____________________

25. (15 days, 14 hours, 36 minutes) ÷ 3

25.____________________

Perform the indicated operations.

26. Change 5:15 a.m. Eastern time to Mountain time.

26.____________________

27. Change 10:50 p.m. Pacific time to Central time.

27.____________________

28. Change 11:00 p.m. to 24-hour time.

28.____________________

29. Change 1530 hours to standard clock time.

29.____________________

30. Change 2300 hours Pacific time to standard clock time in New York City.

30.____________________

31. A plane leaves Buffalo, New York, at 9:25 a.m. If the travel time to San Diego, California, is 6 hours, 15 minutes, what will be the time in San Diego when the plane arrives?

31.____________________

32. A car travels at an average speed (including rest stops) of 50 miles per hour for a trip of 1000 miles. The trip got underway at 10 a.m. Mountain time on Monday. It ended in the Eastern time zone. What was the day and time of arrival?

32.____________________

33. A plane leaves Las Vegas, Nevada, at 1:00 p.m. and arrives in Newark, New Jesey, at 9:35 p.m. What was the travel time?

33.____________________

Chapter 8

PERIMETER, AREA, AND VOLUME

Test Score ______________

Name ______________________________

Date ______________ Class ______________

Complete each of the following problems. Show your work and place answers on the lines provided. Answer *must* contain proper units. (To figure the answer, it might be helpful to make a sketch of the problem.) This is an open-book test.

1. Find the perimeter of a rectangle with sides measuring 150 feet and 225 feet. 1.______________

2. Calculate the length of wood necessary to make a 1-inch-wide frame for an 8-inch by 10-inch picture. 2.______________

3. Determine the length of fencing required to enclose a square playground, 100 meters wide. 3.______________

4. If a square has a perimeter of 60 yards, what is the length of one side? 4.________________

5. A triangle has three, equal-length sides of 128 inches each. What is the perimeter, in *feet?* 5.________________

6. An octagon-shaped (eight-sided) lamp shade needs a border of trim. If each of the eight sides (along the perimeter) is 15 centimeter, how long will be trim be? 6.________________

7. Determine the perimeter of a circle with a radius of 15 feet. 7.________________

8. Find the perimeter of a circle with a diameter of 2.5 meters. 8.______

9. How much area can be covered by a 4-foot by 8-foot sheet of plywood? 9.______

10. What is the total area of three walls, each 8 feet high; two are 14 feet long and the other is 10 feet long? 10.______

11. If a roll of wallpaper is 24 inches wide and 25 yards long, how many rolls should be bought to cover 4 walls, each 8 feet high and 12 feet long? 11.______

12. Carpeting sold by the squard yard will be used to cover an 8-foot by 12-foot floor. What will be the total cost of the carpet if it sells for $2 a yard?

12.____________________

13. A flat roof is 50 feet long and 28 feet wide. What is the area to be covered by roofing material?

13.____________________

14. How many 4-foot by 8-foot sheets of plywood will be needed to cover a flat roof that is 48 feet long and 24 feet wide?

14.____________________

15. As a rough estimate, the cost to build a house can be based on square footage. At $50 per square foot, what would a house cost if it was 52 feet long and 36 feet wide?

15.____________________

16. A house gable, the triangular area under the eaves, is painted with a paint that covers roughly 200 square feet per gallon. If the gable is 32 feet across (at the base) and 10 feet high, how many gallons must be bought?

16.____________________

17. What is the area of a circle with a radius of 24 inches?

17.____________________

18. What is the area of a circle with a diameter of 18 centimeters?

18.____________________

19. Cubes measuring 1 inch on each side are to be packed in a cardboard box. If the box measures 14 inches wide, 16 inches long, and 12 inches high, how many of the cubes can be packed inside?

19.____________________

20. What volume of space does a room measuring 10 feet by 16 feet, with 7 foot ceilings occupy?

20.________________

21. A portable heater blows out heated air at a rate of 2 cubic feet per minute. About how long will it take to fill the space of a triangular tent with an 8-foot by 4-foot floor and a 4-foot peak with heated air?

21.________________

22. Find the volume of a 250-foot-deep water well that has a casing I.D. of 8 inches. (Casing is a metal pipe used to line a well.)

22.________________

23. Cement sold by the cubic yard is to be used to make a footing for a foundation wall. The formwork for the footing is 6 inches high, 24 inches wide, and 16 feet long. How much cement will be needed?

23.________________

24. One cord of firewood is 128 cubic feet. How many cords of wood are in a stack 50 feet long, 4 feet wide, and 6 feet high?

24. ______________________

25. If the water in a bathtub is 8 inches deep, determine how many gallons of water are in the tub if the dimensions are 4 feet by 2 feet. (Ignore tub slope and curvature and round answer to nearest one.)

25. ______________________

TEST • TEST • TEST • TEST • TEST • TEST • TES

Chapter 9

READING TOOLS OF MEASUREMENT

Test Score ______________

Name ______________________________

Date ______________ Class ______________

Complete each of the following problems. Show your work and place answers on the line provided. Use the drawing shown with each problem. Include units of measure with the answer.

1. What are the measurements at points *a, b,* and *c*? (feet/inches from zero)

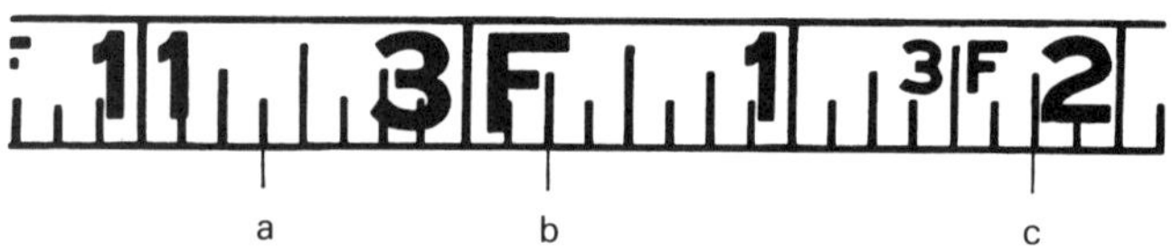

a = ______________

b = ______________

c = ______________

2. What are the measurements at points *a, b,* and *c*? (inches from zero)

a = ______________

b = ______________

c = ______________

3. What are the measurements at points *a, b,* and *c*? (inches from zero)

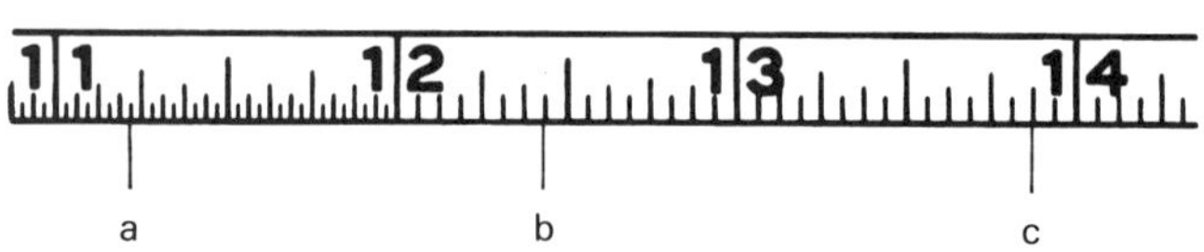

a = ______________

b = ______________

c = ______________

4. What are the measurements at points *a, b,* and *c*? (decimal inches from zero)

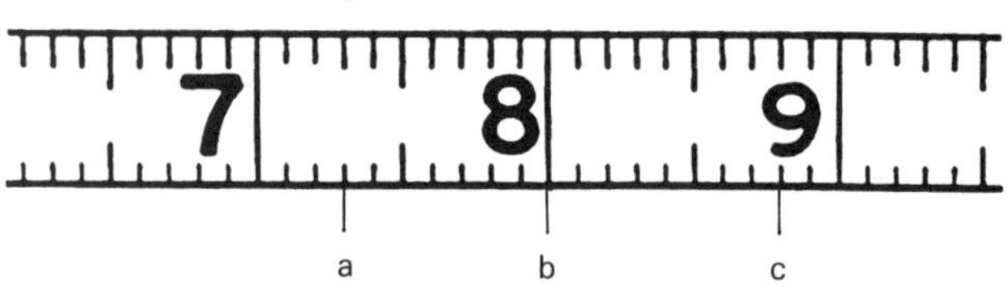

a = ______________

b = ______________

c = ______________

5. What are the measurements at points *a, b,* and *c*? (decimal feet from zero)

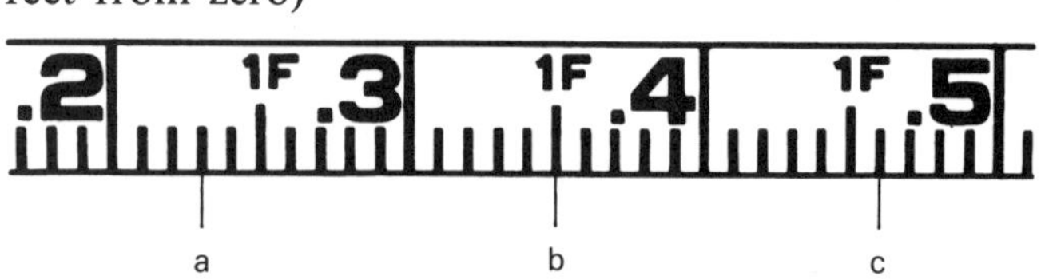

a = ______________

b = ______________

c = ______________

6. What are the measurements of *a, b,* and *c*? (centimeters)

a = ____________

b = ____________

c = ____________

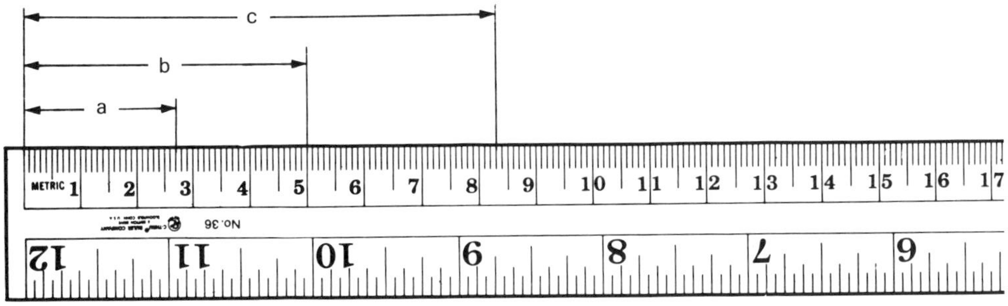

7. What are the measurements of *a, b,* and *c*? (*a:* centimeters; *b* and *c:* millimeters)

a = ____________

b = ____________

c = ____________

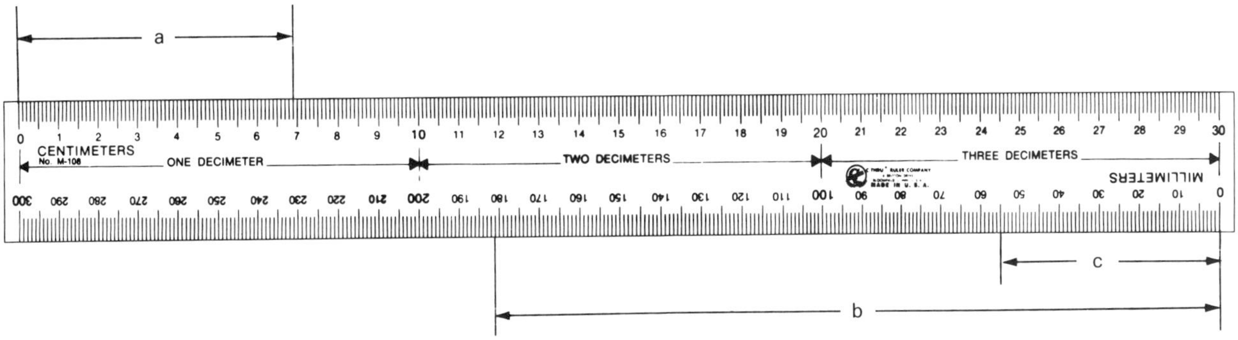

8. What are the measurements of *a, b,* and *c*? (*a:* inches; *b* and c: feet/inches)

a = ____________

b = ____________

c = ____________

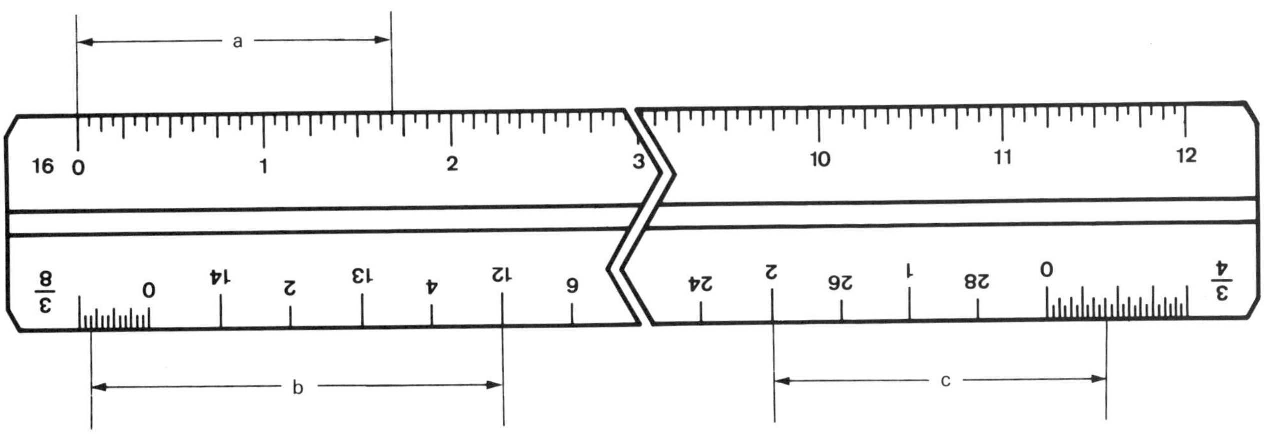

9. What are the measurements of *a, b,* and *c*? (feet/inches)

a = ______________

b = ______________

c = ______________

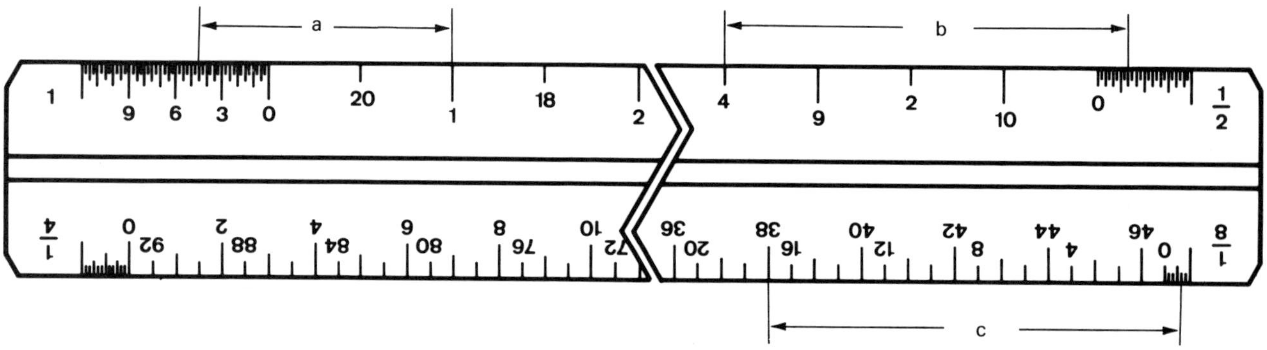

10. What are the measurements of *a, b,* and *c*? (feet)

a = ______________

b = ______________

c = ______________

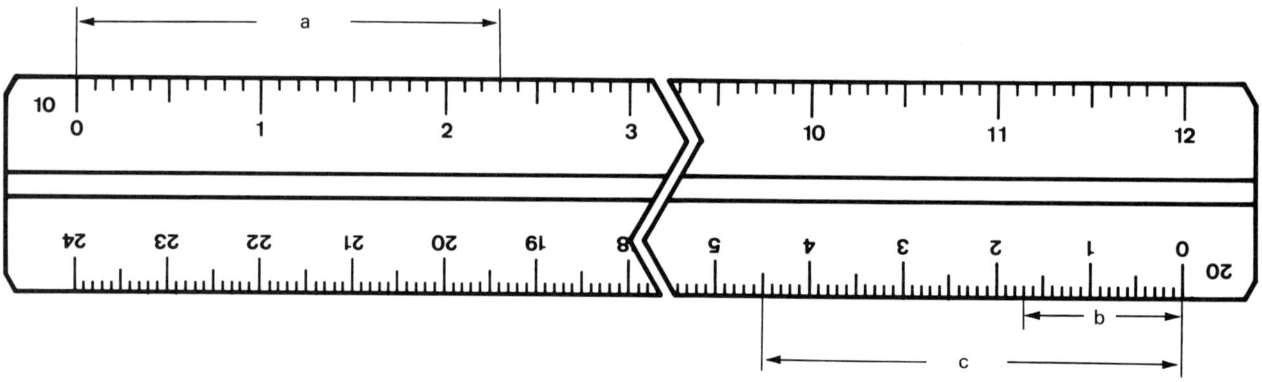

11. What are the measurements of *a, b,* and *c*? (feet/inches)

a = ______________

b = ______________

c = ______________

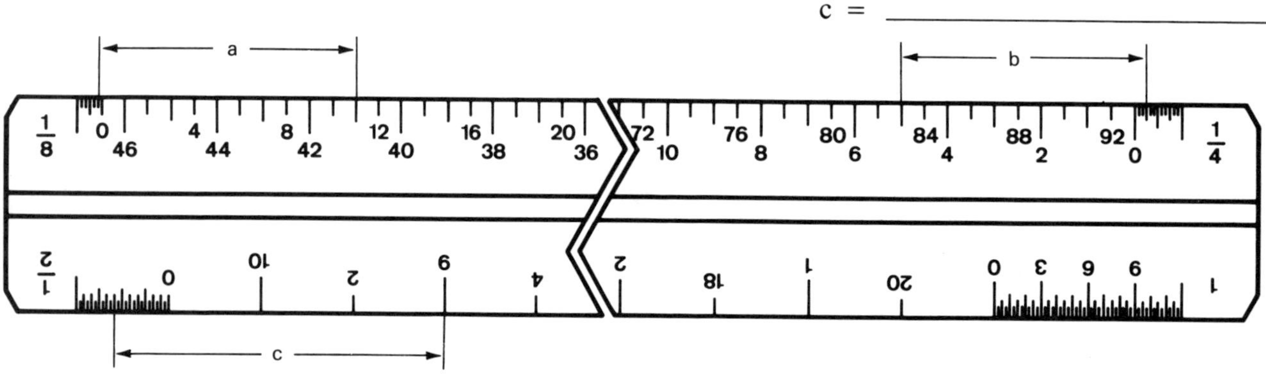

12. What measurement is read on the micrometer scale? (inches)

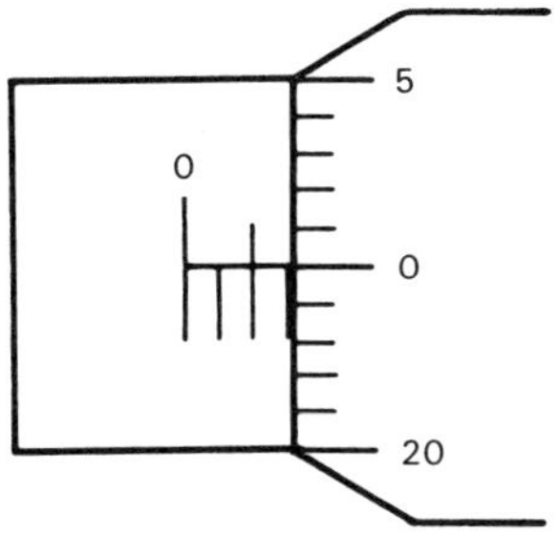

Sleeve = ____________________

Thimble = ____________________

Measurement = ____________________

13. What measurement is read on the micrometer scale? (inches)

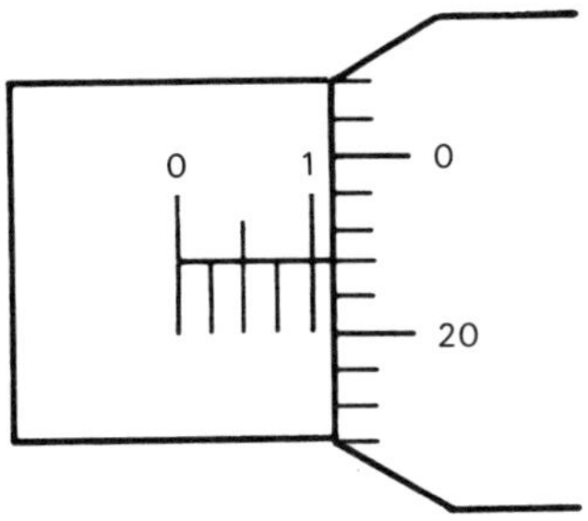

Sleeve = ____________________

Thimble = ____________________

Measurement = ____________________

14. What measurement is read on the micrometer scale? (inches)

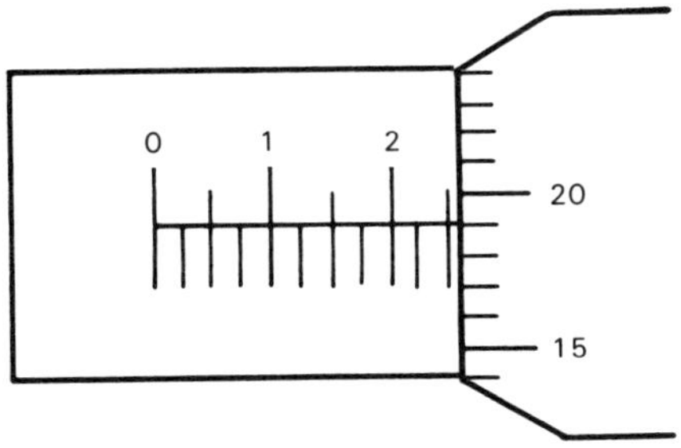

Sleeve = ____________________

Thimble = ____________________

Measurement = ____________________

15. What measurement is read on the micrometer scale? (inches)

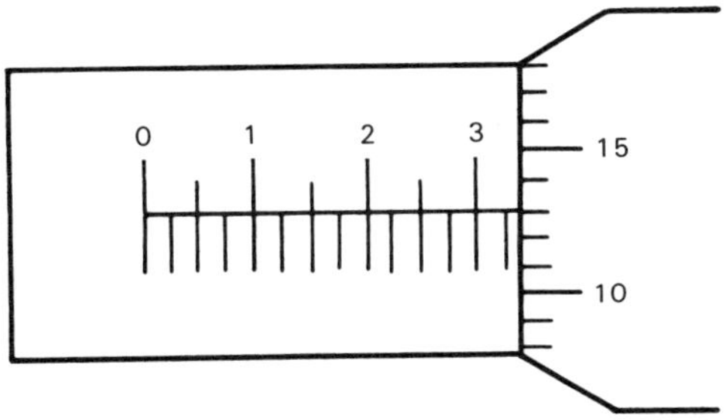

Sleeve = ____________________

Thimble = ____________________

Measurement = ____________________

16. What measurement is read on the micrometer scale? (inches)

Sleeve = ____________

Thimble = ____________

Measurement = ____________

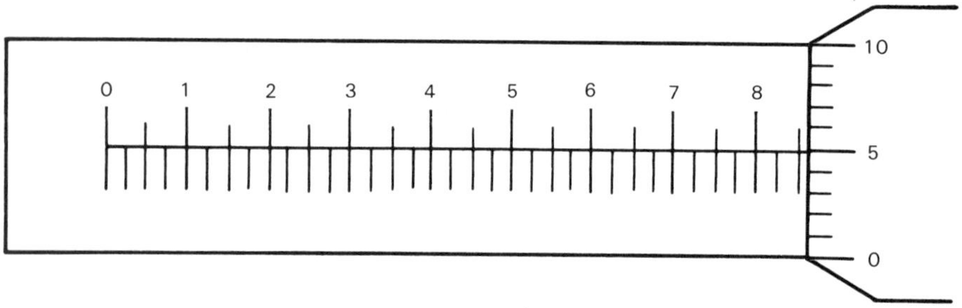

17. What measurement is read on the micrometer scale? (millimeters)

Sleeve = ____________

Thimble = ____________

Measurement = ____________

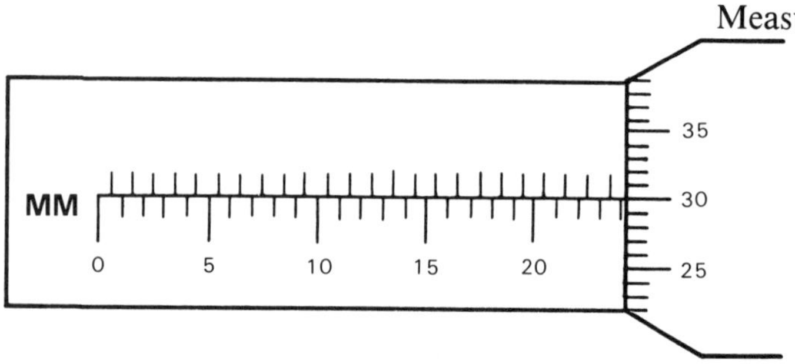

18. What measurement is read on the micrometer scale? (millimeters)

Sleeve = ____________

Thimble = ____________

Measurement = ____________

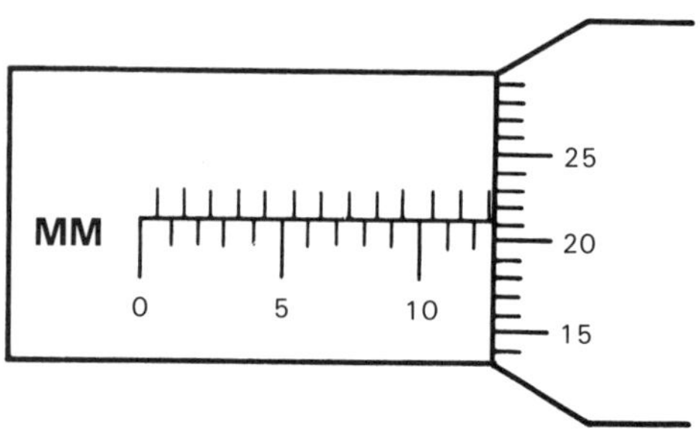

19. What measurement is read on the micrometer scale? (millimeters)

Sleeve = ____________________

Thimble = ____________________

Measurement = ____________________

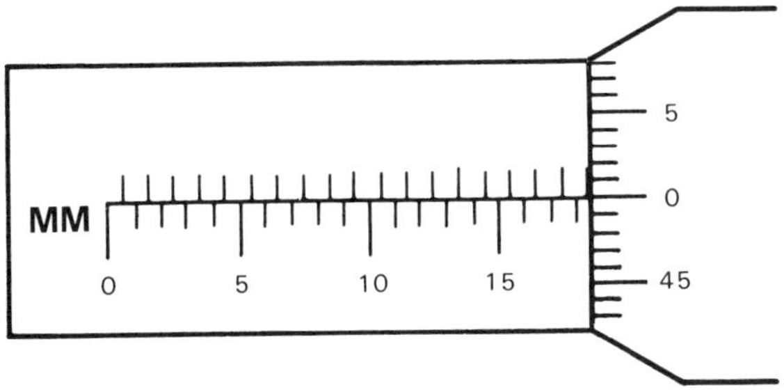

20. What measurement is read on the micrometer scale? (millimeters)

Sleeve = ____________________

Thimble = ____________________

Measurement = ____________________

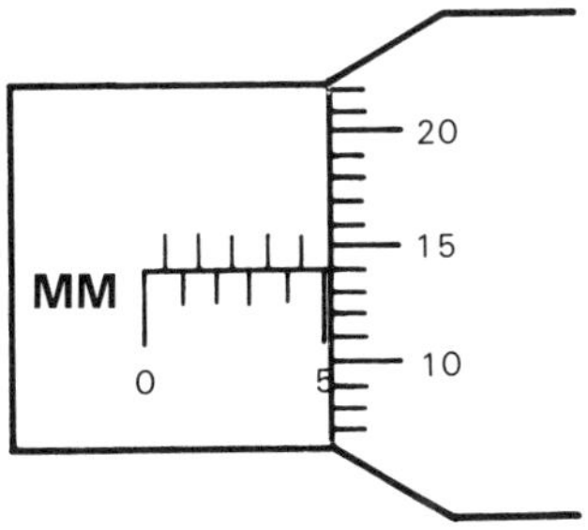

21. What measurement is read on the micrometer scale? (millimeters)

Sleeve = ____________________

Thimble = ____________________

Measurement = ____________________

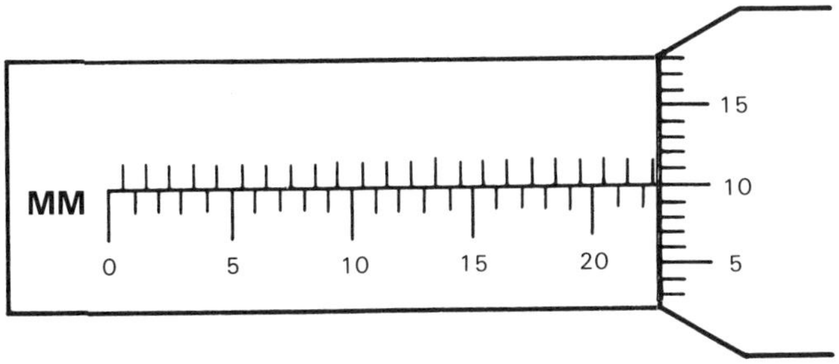

22. What measurement is read on the micrometer scale? (inches)

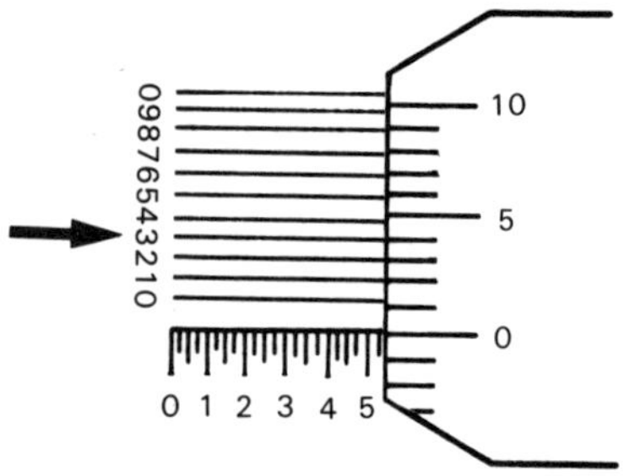

Sleeve = ____________

Thimble = ____________

Vernier = ____________

Measurement = ____________

23. What measurement is read on the micrometer scale? (inches)

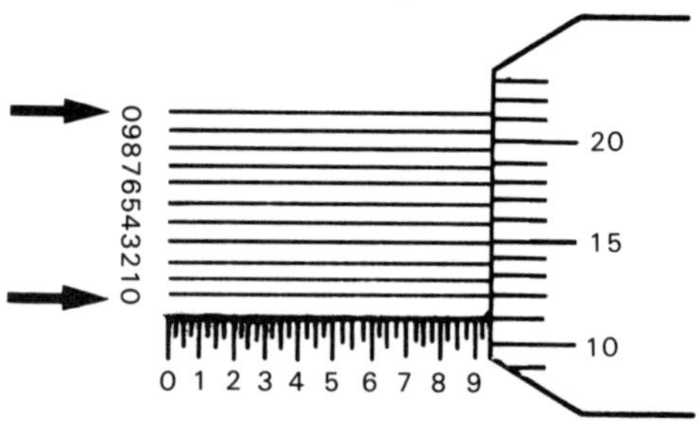

Sleeve = ____________

Thimble = ____________

Vernier = ____________

Measurement = ____________

24. What measurement is read on the micrometer scale? (inches)

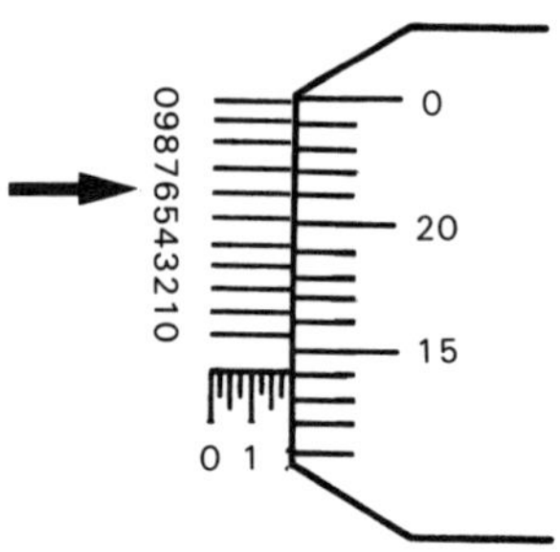

Sleeve = ____________

Thimble = ____________

Vernier = ____________

Measurement = ____________

25. What measurement is read on the micrometer scale? (inches)

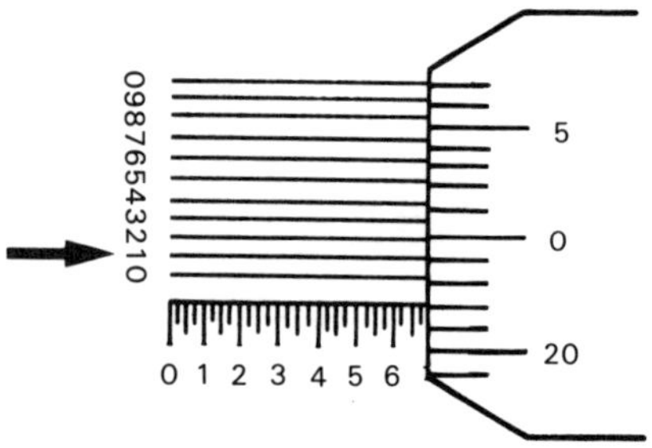

Sleeve = ____________

Thimble = ____________

Vernier = ____________

Measurement = ____________

For Questions 26 through 33, "range" is the position of the range switch: "scale" is the numbers across the scale, with the decimal adjusted to the range; "line" is the value of each minor line of the adjusted scale.

26. With the multimeter shown: range = ________

scale = 0, ______, ______, ______, ______, ______

line = ________

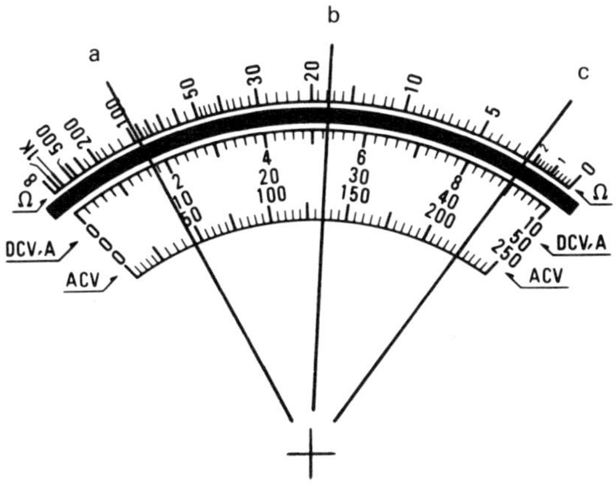

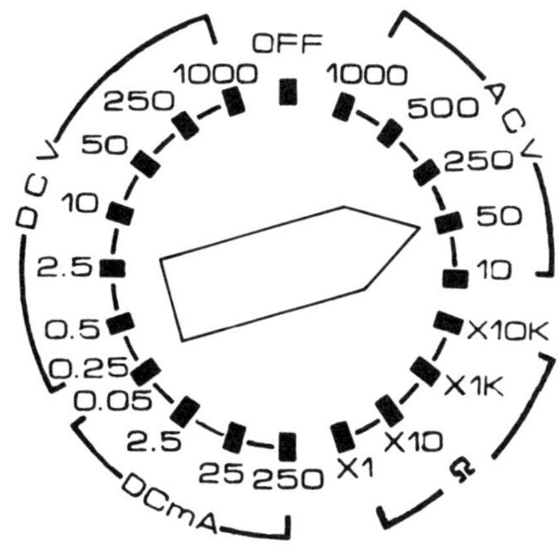

27. Read the meter needles at *a,* then *b,* then *c.*

a = ______________________

b = ______________________

c = ______________________

28. With the multimeter shown, fill in the table.

NEEDLE	SMALL LINE	READING	RANGE	MEASUREMENT
a				
b				
c				

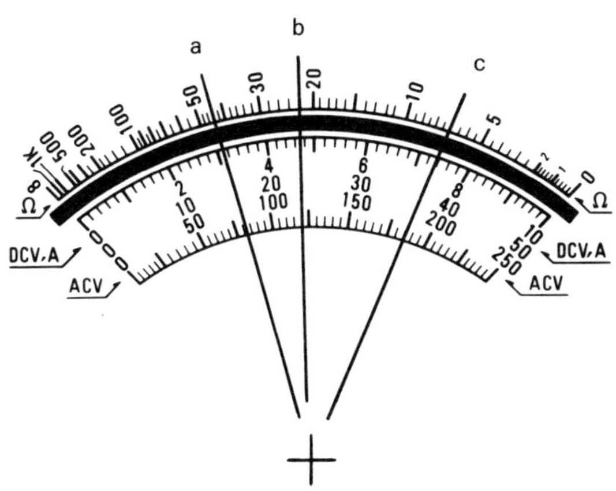

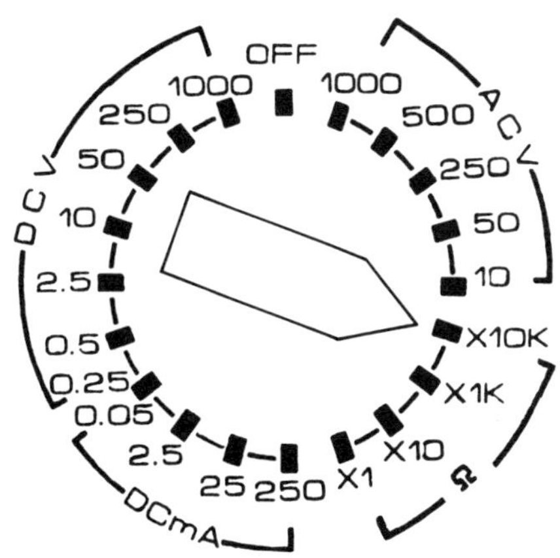

29. With the multimeter shown:

range = ________

scale = 0, ______, ______, ______, ______, ______

line = ________

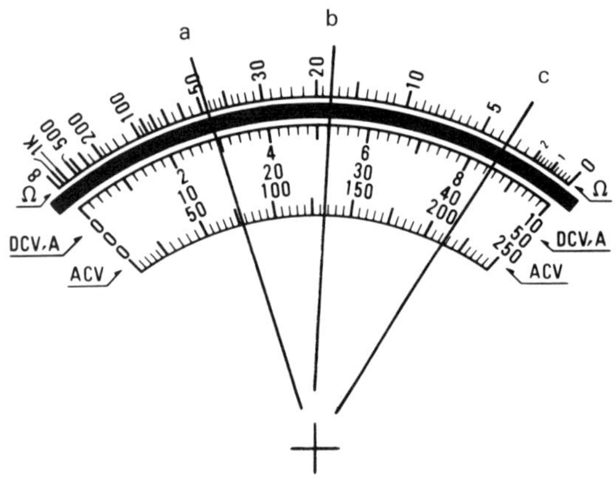

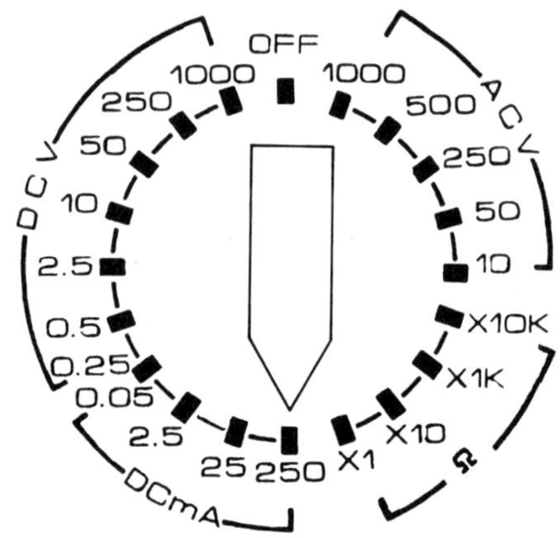

30. Read the meter needle at *a*, then *b*, then *c*.

a = ____________________

b = ____________________

c = ____________________

31. With the multimeter shown:

range = ________

scale = 0, ______, ______, ______, ______, ______

line = ________

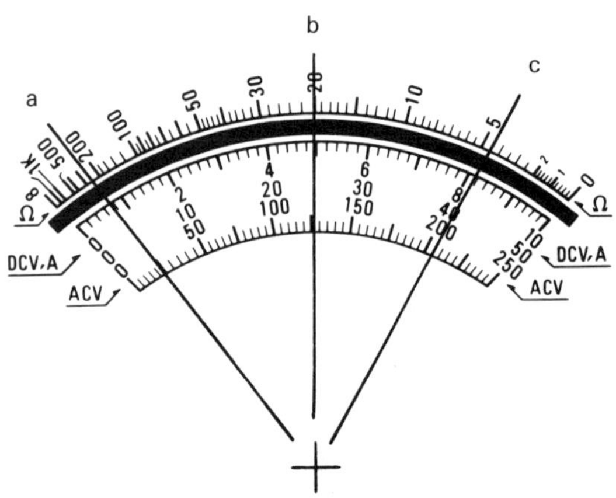

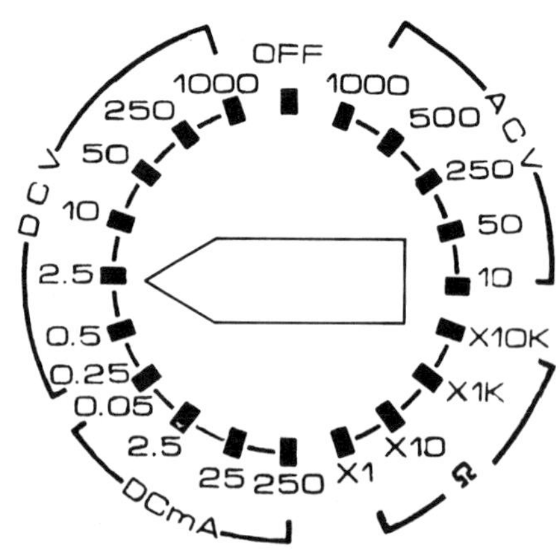

32. Read the meter needle at *a*, then *b*, then *c*.

a = ____________________

b = ____________________

c = ____________________

Chapter 10

SIGNED NUMBERS AND OPERATIONS

Test Score ______________

Name ______________________________

Date ______________ Class ______________

Complete each of the following problems. Show your work and place answers on the lines provided.

Show each of the following figures as signed numbers. It is not necessary to show units.

1. A \$25 purchase. 1. ______________
2. A savings account of \$560. 2. ______________
3. Normal body temperature is 98.6° Fahrenheit. 3. ______________
4. Ground freezes to 3 feet below the surface. 4. ______________
5. On a test, a student has 5 wrong answers. 5. ______________

Give the absolute value of the following:

6. $|-10|$ = ______________   7. $|2.2|$ = ______________

Insert the correct inequality sign (> or <) in the blank.

8. 6 ______________ −9
9. −2 ______________ −8
10. 0 ______________ (−4 − 9)

Indicate if these fractions are positive (+) or negative (−).

11. $-\frac{2}{-3}$ ⇒ ________   12. $-\frac{-5}{6}$ ⇒ ________

13. $\frac{-1}{4}$ ⇒ ________   14. $\frac{4}{-7}$ ⇒ ________

15. $-\frac{-1}{-2}$ ⇒ ________   16. $\frac{-3}{-5}$ ⇒ ________

Perform the indicated operations.

17. $6 - 8$

17.________________

18. $9 - 3$

18.________________

19. $-5 - 7$

19.________________

20. $-8 - 2$

20.________________

21. $4 + (-3)$

21. ______________________

22. $(-1) + 6$

22. ______________________

23. $-10 - -1$

23. ______________________

24. $-2 - -14$

24. ______________________

25. $(-2) + (-5) - (-9) + 21 - 6 - 13$

25.____________________

26. $\frac{1}{2} - \frac{3}{4} + \frac{1}{6} - \frac{2}{3} + \frac{1}{4}$

26.____________________

27. 5×-6

27.____________________

28. $-3 \times -2 \times -1$

28.____________________

29. $-\frac{2}{3} \times \frac{3}{5}$

29. ______________________________

30. $-\frac{1}{9} \times -\frac{5}{7}$

30. ______________________________

31. $\frac{-24}{-8}$

31. ______________________________

32. $-\frac{-16}{-4}$

32. ______________________________

33. $-3\overline{)21}$

33.____________________

34. $6\overline{)-60}$

34.____________________

35. $-2.2 \div 11$

35.____________________

36. $4 \div -20$

36.____________________

Find the solution.

37. $(3 - 7 + 5) - [-4 + (1 - 2)]$ 37.________

38. $-4(-2) + 16 - (3 - 8) - 1(0)$ 38.________

39. $(5 - 6) - (1)(-9) - (-7 - 3)$ 39.________

40. $[-2 - (3 + 1 - 4) + 2]$ 40.________

41. 7^2 = ________________

42. $(-4)^3$ = ________________

43. $\sqrt{49}$ = ________________

44. $\sqrt[3]{-27}$ = ________________

Approximate the square root to 2 significant figures.

45. $\sqrt{60}$ = ________________

46. $\sqrt{90}$ = ________________

Find the solution.

47. $-[3(-5) + 2(-4 - 9)]$

47.____________________

48. $-2(3 + 1) + 4 \div -2 - (-8 + 2)$

48. ______________________

49. $\dfrac{2 + \sqrt{36} - 4(-1)}{0 - 7(-2) + -8}$

49. ______________________

APPLICATIONS OF SIGNED NUMBERS

50. After a 50 pound box of nails was opened, three carpenters each took out 5 pounds. Before the end of the day: the first carpenter returned 2 pounds; the second returned 3 pounds; the third took an extra pound. The second day, they each took out 7 pounds. At the end of the day, they emptied their pouches. The first carpenter returned 1 pound; the second returned 4 pounds; the third returned 2 pounds. How many pounds of nails were used on the job?

50. ______________________

Chapter 11

INTRODUCTION TO ALGEBRA

Test Score ______________

Name ______________________________

Date ______________ Class ______________

Complete each of the following problems. Show your work and place answers on the lines provided.

With the algebraic expression given, write each term. Also, identify the coefficients, variables, constants and constant terms. Include the + or − signs.

		Term	Coefficient	Variables	Constant	Constant Term
Example: $x^2 + 2xy - 14$		$+x^2$	$+1$	x	$+1$	
		$+2xy$	$+2$	x,y	$+2$	
		-14			-14	-14
1. $4x^2 - 3 + xy$	1.					
2. $6 - 5uv + u^2$	2.					
3. $-10 - 3x^2y^2 + 2xy$	3.					
4. $(4)(3x) - y^2 + 1$	4.					
5. $6x^2 - 3xyz - 13$	5.					

Write each term, then separate the factors. Only factor if multiplication is indicated. Include + or − sign.

		Term	Factors
Example: $3x^2 - 4(x + 1) - 32$		$+3x^2$	$+3 \cdot x \cdot x$
		$-4(x + 1)$	$-4 \cdot (x + 1)$
		-32	
6. $7 + 4a^2b - 3ab^2$	6.		
7. $-5(x + 1) - 3y + 6xyz$	7.		
8. $(-3)(-2x) - 4x^2 + 2^3$	8.		
9. $5 + (3x)(xy^2) - 10y$	9.		
10. $4^2 + x^3y - (6x)(7x)$	10.		

Perform the indicated arithmetic operations. Simplify all answers.

11. $3x - 6x + 5 - 7x + 8$ 11. ______________

12. $a(5a + c) - 2a(c - 4a)$

12.____________________

13. $x^0 \cdot 2x^1 \cdot 4x^2 \cdot 2x^3$

13.____________________

14. $(a^3b^4)^2$

14.____________________

15. $\frac{x^2(y + 3x) \cdot 6y}{12y^2(3x + y) \cdot x}$

15.____________________

16. $(6x + 5) - (2x - 8)$

16.____________________

17. $4(-2x - 5) - 2(x + 6)$

17.____________________

18. $(8y - 10x) - 2x(3 - y) + 3y(x + y - 5)$

18.____________________

19. $2a^6b^4c^2 \cdot 2a^3b^5c^7$

19.____________________

20. $3ab(-a + 5) - 4a^3b + 5(ab - 2)$ 20.________________

21. $(x^2 + 5x - 14) \div (x - 2)$ 21.________________

Factor each of the following expressions completely.

22. $2x^2 - 30x + 108$ 22.________________

23. $25x^2 - 36y^2$ 23.________________

Simplify each of the following fractions.

24. $\dfrac{x^2(x - 9)}{x^2 - 81}$

24. ______________________

25. $\dfrac{x^4 - 81}{x^2 + x - 6}$

25. ______________________

Chapter 12

ALGEBRAIC EQUATIONS AND INEQUALITIES

Test Score ______________

Name ______________________________

Date ______________ Class ______________

Complete each of the following problems. Show your work and place answers on the lines provided.

Solve the following equations.

1. $a + 10 = -3$ 1.______________

2. $9 - x = 12$ 2.______________

3. $b - 2 = 8 - b$ 3.______________

4. $4x + 3x = 28$ 4.______________

5. $7y - 8 = 4y + 10$ 5.______________________________

6. $12d + 4 = -8$ 6.______________________________

7. $\frac{x}{4} = -8$ 7.______________________________

8. $\frac{24}{3x} = 8$ 8.______________________________

9. $8(a + 2) = 7(1 + a)$ 9.______________________________

10. $x(7 - 2) = 2(7 - x)$ 10. ____________

11. $(8 + x) - 6 = -1(3x + 2)$ 11. ____________

12. $\frac{3}{4} + \frac{x}{3} = \frac{x}{4}$ 12. ____________

13. $3.9a - 0.9a + 1.0 = 2a$ 13. ____________

14. $3x^2 = 27$ 14. ____________

15. $a^2 = 64$ 15. ____________

16. $3x^2 + 5 = 152$

16. ______________________

17. $7 + \frac{x^2}{3} = 34$

17. ______________________

18. $2(x + 3) = 62$

18. ______________________

Solve the following inequalities.

19. $a - 5 < 13$

19. ______________________

20. $5b - 7 \geq 8$

20. ______________________

Solve the systems of equations using the addition/subtraction method.

21. $5b - c = 33$
$7b + c = 51$

21. b = ______ c = ______

22. $2x + y = 13$
$5x + y = 28$

22. x = ______ y = ______

Solve the systems of equations using the substitution method.

23. $2x + y = 0$
$3x - y = 10$

23. x = ______ y = ______

24. $y - x = -2$
$x + 5y = 20$

24. x = ______ y = ______

Chapter 13

GRAPHS OF LINEAR EQUATIONS AND INEQUALITIES

Test Score ____________

Name ____________

Date ____________ Class ____________

Complete each of the following problems. Show your work and place answers on the lines provided.

For questions 1 through 8 refer to the graph shown.

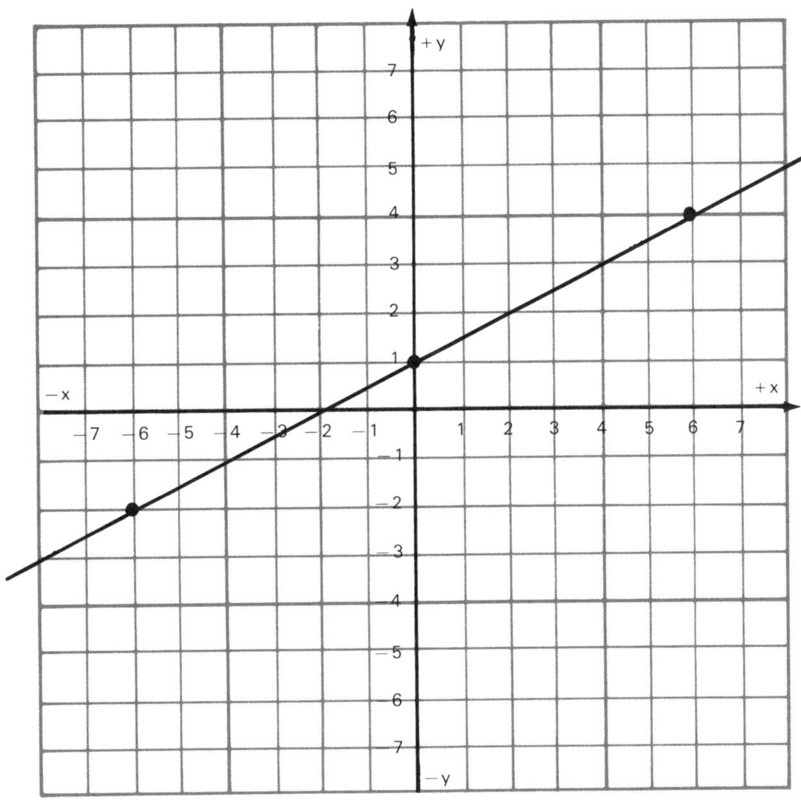

1. What are the coordinates of the origin on this graph? 1.____________
2. This graph is a plot of what type of equation? 2.____________
3. What is the slope of the line? 3.____________
4. Give the equation of this line in slope-intercept form. 4.____________
5. Give the ordered pair found in the third quadrant. 5.____________
6. Does this graph represent a functional relationship? 6.____________
   Explain. ____________

7. What is the value of the independent variable at x = 6? 7.__________

8. For what values of x are y-values positive? 8.__________

9. Use the graph paper provided to solve these simultaneous equations graphically. Circle the solution.

$$y = x + 1$$

$$x + y = 5$$

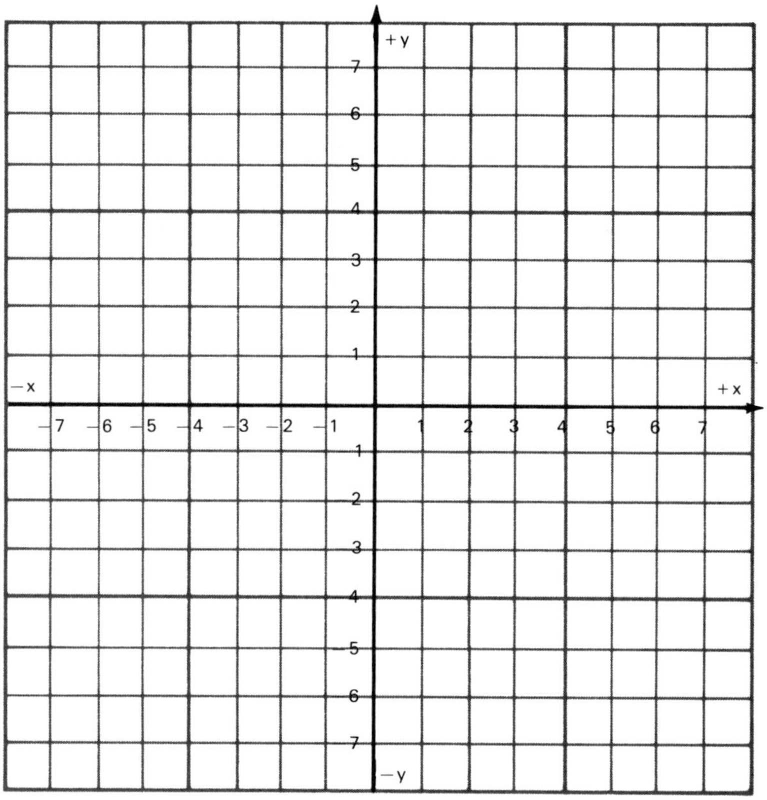

10. What is the solution of the system of equations given in the previous question? 10.__________

Chapter 14

SOLVING PROBLEMS WITH ALGEBRA

Test Score ______________

Name ______________________________

Date ______________ Class ______________

Complete each of the following problems. Show your work and place answers on the lines provided. Include units where necessary.

Use the relationship given to determine the unknown number in each pair.

1. Relationship: In one hour, 125 hamburgers were sold through employees combined efforts at the drive-thru window and the inside counter.

	drive-thru (burgers)	inside (burgers)
a.	75	______
b.	______	90
c.	x	______
d.	______	y + 5
e.	3z	______

Solve the following word problems.

2. *Geometry problem.* If two opposite sides of a square are each increased 12 inches, and the other two sides are each increased 4 inches, a rectangle is formed having a perimeter double that of the square. How long are the sides of the square?

2. ______________________

3. *Motion problem.* A single-engine plane leaves from the airport and flies at an average airspeed of 75 mph. Three hours later, a faster, single-engine plane leaves from the same airport and flies at an average airspeed of 100 mph. How long will it take the faster plane to overtake the slower plane?

3.____________________

4. A 40-foot rope is cut into two pieces in such a way that one piece is 4 feet longer than twice the other. How long is each piece?

4.____________________

____

5. Find three consecutive odd integers such that the first plus twice the second plus three times the third is equal to 70.

5. ______________________

____

____

6. *Mixture problem.* A farmer sold 100 bushels of apples to a wholesale dealer. One type of apple was 90¢ per bushel, and a different type was 50¢ per bushel. If the farmer received $78.80 for the total sale, how many bushels of each were sold?

6. ______________________

____

7. *Mixture problem.* There are 10 quarts of a 20% antifreeze solution in the partially filled radiator of a car. How many quarts of antifreeze must be added to make a 50% antifreeze solution?

7.______________________

8. *"Work" problem.* Paychecks of a certain company are printed by a new computer in 3 hours, where the old computer takes 9 hours. What is the total time to print the paychecks with the computers operating together?

8.______________________

Solve the following problems of ratio, proportion, and variation.

9. Two quarts (64 ounces) of film developing solution are to be mixed. The instructions state to mix 2 parts of concentrated developer to 6 parts water. Determine how many ounces of each should be mixed.

9. ______________________

____

10. A building casts a shadow 90 feet long, and a 4-foot post casts a shadow 6 feet long. What is the height of the building?

10. ______________________

11. Find y when x = 10 if y varies directly as x and y = 20 when x = 8.

11.______________________

12. Find y for x = 6 and z = 5 if y varies directly as x and inversely as z and y = 120 when x = 4 and z = 10.

12.______________________

Chapter 15

APPLIED TRIGONOMETRY

Test Score ______________

Name ______________________________

Date ______________ Class ______________

Complete each of the following problems. Show your work and place answers on the lines provided. Include units where necessary.

Use trigonometric functions and the Pythagorean theorem to determine the requested measurements. Round answers to one decimal place, where necessary.

1. a. length of rafter a — 1a. ______________
   b. length of rafter b — b. ______________

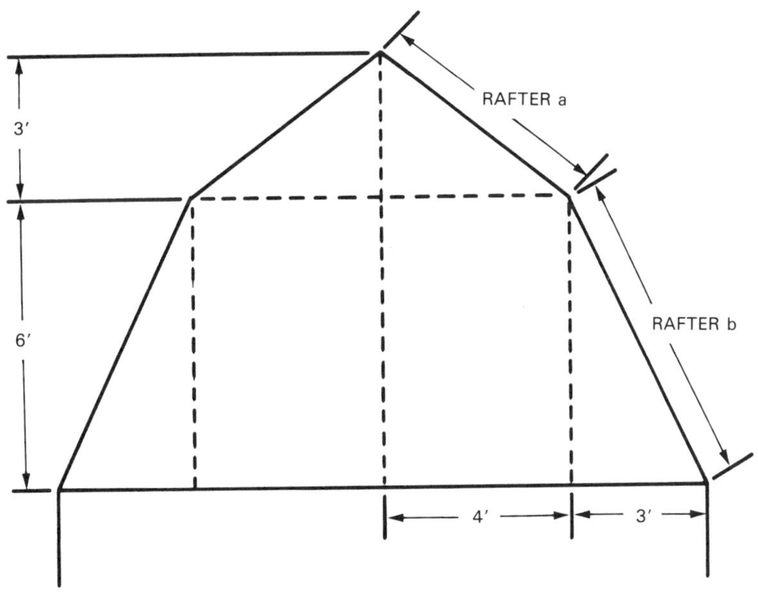

2. a. height of antenna
   b. length of guy wire
   c. angle B

2a. ______
b. ______
c. ______

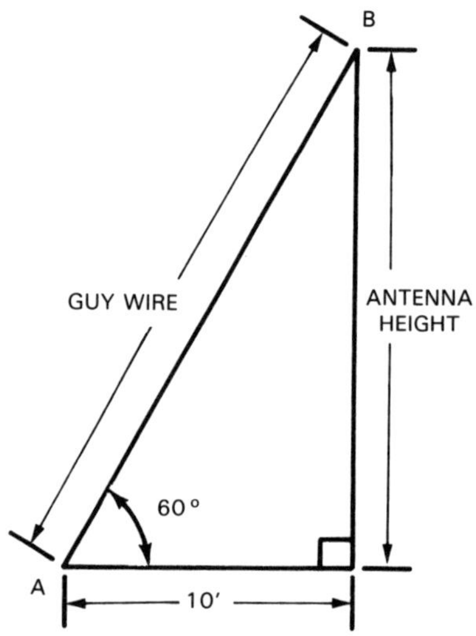

3. a. length of stairway stringer
   b. angle A
   c. angle B

3a. ______
b. ______
c. ______

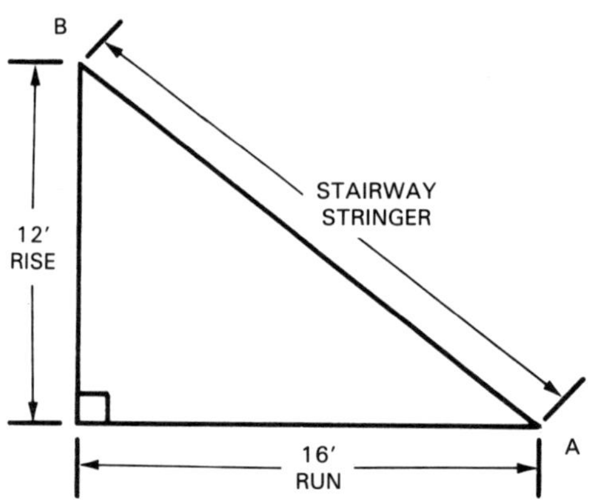

4. a. length of vertical support
   b. length of base support
   c. angle B

4a. ______
b. ______
c. ______

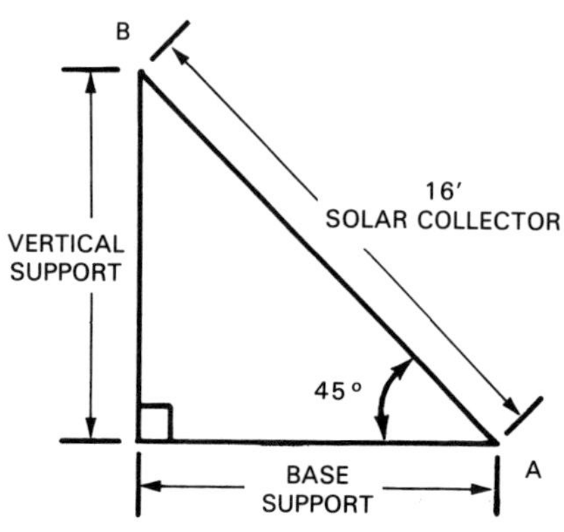

Use the Law of Sines or the Law of Cosines, as needed, to solve the following triangles (find the unknown dimensions requested). Note that the order of solving does not necessarily follow the order in which the unknowns are listed. Round all answers to two decimal places, where necessary. The formulas are provided here for reference.

Law of Sines:

$$\frac{a}{\sin A} = \frac{b}{\sin B} = \frac{c}{\sin C}$$

Law of Cosines:

$$a^2 = b^2 + c^2 - 2bc\,(\cos A)$$

$$b^2 = a^2 + c^2 - 2ac\,(\cos B)$$

$$c^2 = a^2 + b^2 - 2ab\,(\cos C)$$

5. length of northerly boundary ____________ 5.
   length of westerly boundary ____________
   angle B ____________

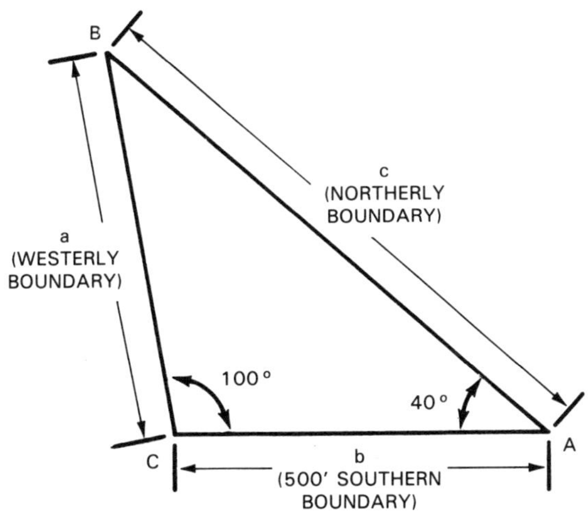

6. length of base ____________ 6.
   angle B ____________

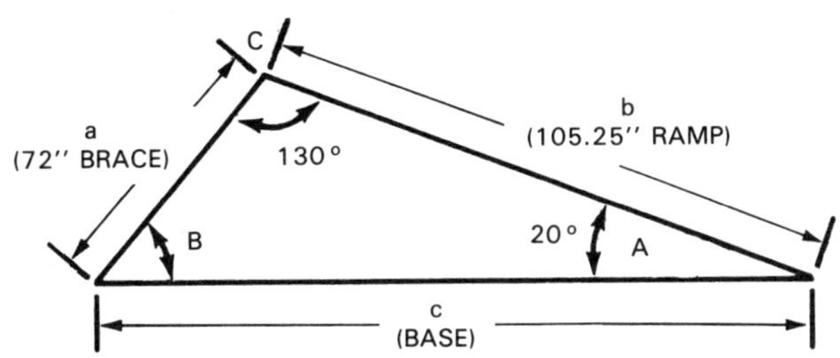

7. measure of length A
   angle B
   angle C

7.__________
____
____

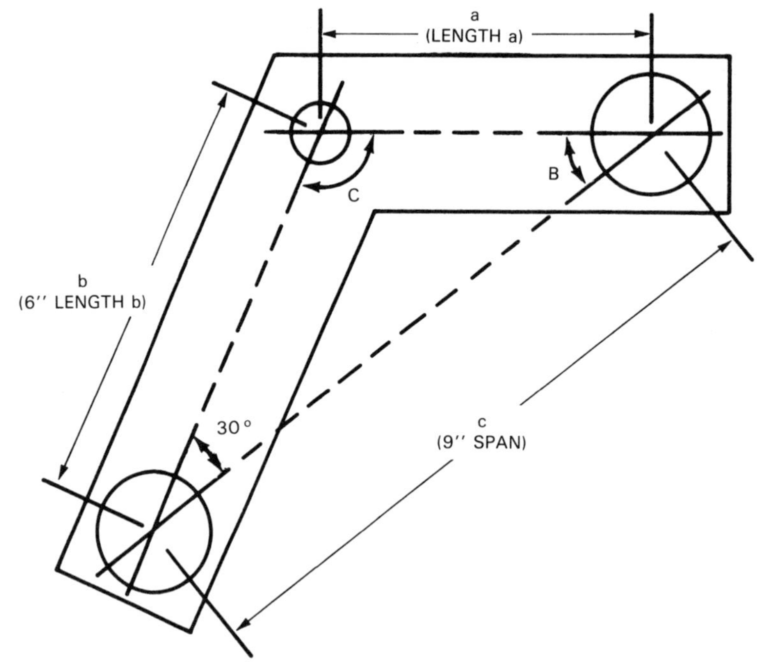

8. angle A
   angle B
   angle C

8.__________
____
____

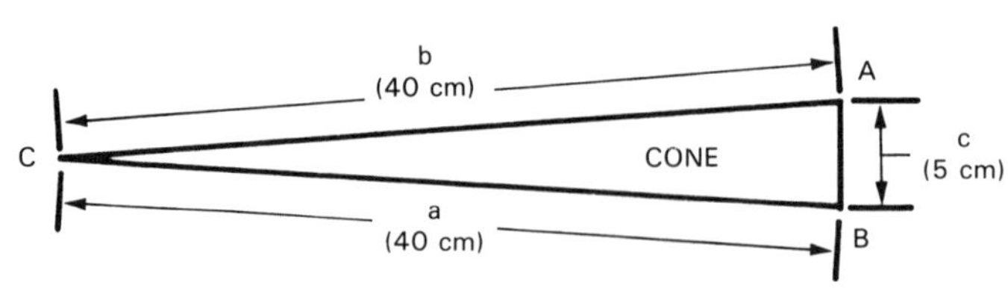

Chapter 16

SCIENTIFIC AND ENGINEERING NOTATION

Test Score ______________

Name ______________________________

Date ______________ Class ______________

Complete each of the following problems. Show your work and place answers on the lines provided. Include units where necessary.

Write the following numbers in the proper form of scientific notation. The decimal point should follow the first significant figure.

Example: 56,000 ⇒ 5.6×10^4

1. 1,280,000 — 1. ______________
2. 535 — 2. ______________
3. 4 — 3. ______________
4. 0.66 — 4. ______________
5. 0.00091 — 5. ______________

Adjust the decimal point and exponent to write the number in the proper form of scientific notation. The decimal should follow the first significant figure.

Example: 56.0×10^3 ⇒ 5.6×10^4

6. 250×10^3 — 6. ______________
7. 7500×10^{-3} — 7. ______________
8. 0.0032×10^5 — 8. ______________
9. 0.00056×10^2 — 9. ______________
10. 0.94×10^{-5} — 10. ______________

Change the following numbers to ordinary notation.

Example: $56.0 \times 10^3 \Rightarrow$ 56,000

11. 8.9×10^6 — 11.____________

12. 4.7×10^{-3} — 12.____________

13. 0.028×10^4 — 13.____________

14. 0.0064×10^{-1} — 14.____________

15. 601×10^3 — 15.____________

Write the following numbers in the proper form of E notation. The decimal point should follow the first significant figure and the exponent should have two digits.

Example: 56,000 $\Rightarrow$ 5.6 E04

16. 270,000,000,000 — 16.____________

17. 0.000000000235 — 17.____________

18. 618 — 18.____________

19. 0.0105 — 19.____________

20. 9.1 — 20.____________

Remove the E notation from the following numbers.

Example: 5.6 E04 $\Rightarrow$ 56,000

21. 111 E06 — 21.____________

22. 12.5 E−03 — 22.____________

23. 0.0046 E05 — 23.____________

24. 0.000121 E−01 — 24.____________

25. 0.000057 E04 — 25.____________

Convert the number on the left to a new number having the multiplier shown with the blank lines. If no multiplier is included, it is the basic unit. A table of multipliers and their exponents is provided for reference.

Example: 56 kΩ = 0.056 MΩ = 56,000 Ω (ohms)

26. 34,000 W = ____________ kW = ____________ MW (watts)

27. 1.25 V = ____________ kV = ____________ mV (volts)

28. 0.46 ms = ____________ μs = ____________ s (seconds)

29. 630 kHz = ____________ MHz = ____________ Hz (hertz)

30. 0.05 MΩ = ____________ kΩ = ____________ Ω (ohms)

Multiplier	Symbol	Power of 10
tera	T	10^{12}
giga	G	10^{9}
mega	M	10^{6}
kilo	k	10^{3}
basic unit		10^{0}
milli	m	10^{-3}
micro	μ	10^{-6}
nano	n	10^{-9}
pico	p	10^{-12}

Perform the indicated arithmetic with the following sets of numbers. Write the answer in the proper form of notation used in the problem. Do not round answers.

31. Add: $125 \times 10^{3} + 0.035 \times 10^{5}$ 31.____________________

32. Add: $0.0267 \times 10^{-5} + 40.32 \times 10^{-7}$ 32.____________________

33. Add: 90.3 E01 + 0.12 E03 33.____________________

34. Add: 50.4 milli + 250 micro 34.____________________

35. Add: 0.034 tera + 80.8 giga 35.____________________

36. Subtract: $43 \times 10^{-3} - 0.000081 \times 10^{2}$ 36.____________________

37. Subtract: 0.085 E−02 − 45.6 E−05 37.________________

38. Subtract: 267 E05 − 0.056 E08 38.________________

39. Subtract: 5.6 mega − 235 kilo 39.________________

40. Subtract: 8.01 milli − 756 micro 40.________________

41. Multiply: $50 \times 10^{1} \times 2.5 \times 10^{0}$ 41.________________

42. Multiply: $0.46 \times 10^{-2} \times 0.100 \times 10^{-4}$ 42.________________

43. Multiply: 12.4 E05 × 0.06 E−07 43.________________

44. Multiply: 0.00006 E−03 × 43,000 E04 44.________________

45. Multiply: 350 milli $\times$ 0.02 kilo

45. ______________________

46. Divide: $\dfrac{840 \times 10^4}{0.42 \times 10^3}$

46. ______________________

47. Divide: $\dfrac{15.0 \times 10^{-6}}{0.03 \times 10^{-4}}$

47. ______________________

48. Divide: $\dfrac{0.024\ \text{E}-05}{40\ \text{E}02}$

48. ______________________

49. Divide: $\frac{6.7\ \text{E}07}{0.005\ \text{E}-01}$

49.____________________

50. Divide: $\frac{0.035\ \text{kilo}}{70\ \text{milli}}$

50.____________________

Chapter 17

BASIC RELATIONSHIPS OF ELECTRICITY

Test Score ______________

Name ______________________________

Date ______________ Class ______________

Complete each of the following problems. Show your work and place answers on the lines provided. Include units where necessary.

Use the given values to find the unknown quantity.

1. V = 75 V   I = 2 A   R = ________

2. I = 0.25 A   R = 600 Ω   V = ________

3. V = 120 V   R = 1000 Ω   I = ________

4. P = 48 W   V = 12 V   I = ________

5. P = 60 W V = 120 V R = ______

6. P = 24 mW I = 100 mA R = ______

7. P = 100 W I = 2 A V = ______

8. P = 40 W R = 10 Ω I = ______

9. I = 15 A V = 120 V P = ______

10. I = 5 A R = 30 Ω P = ______

Use the information available on the schematic diagrams to find the unknown values. Give answers in proper engineering notation and to no more than three significant figures.

11.

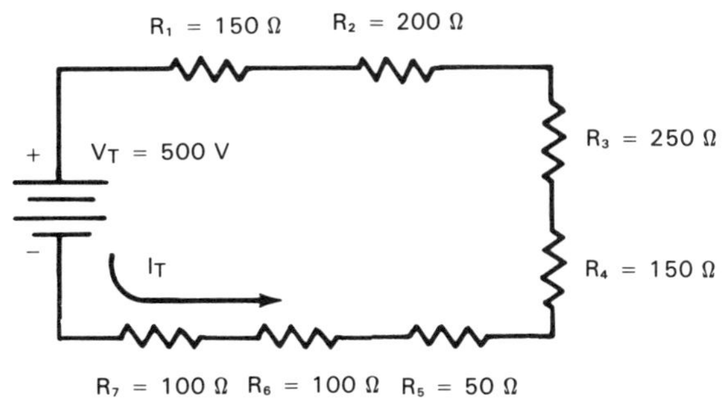

a. Total resistance (R_T) 11a. ____________

b. Total current (I_T) b. ____________

c. Total power (P_T) c. ____________

12.

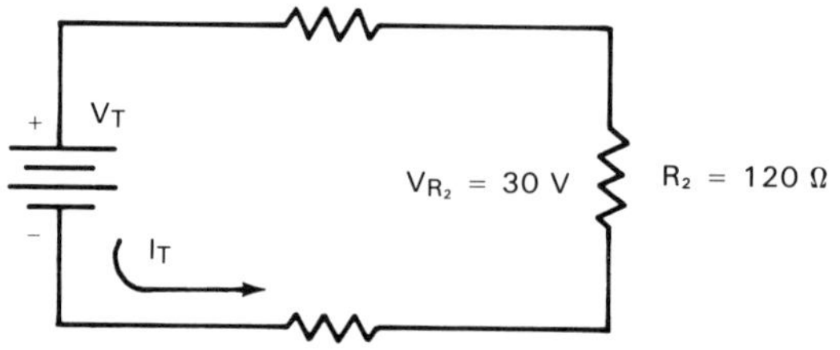

a. Total resistance (R_T) 12a. ______________________

b. Total current (I_T) b. ______________________

c. Applied circuit voltage (V_T) c. ______________________

13.

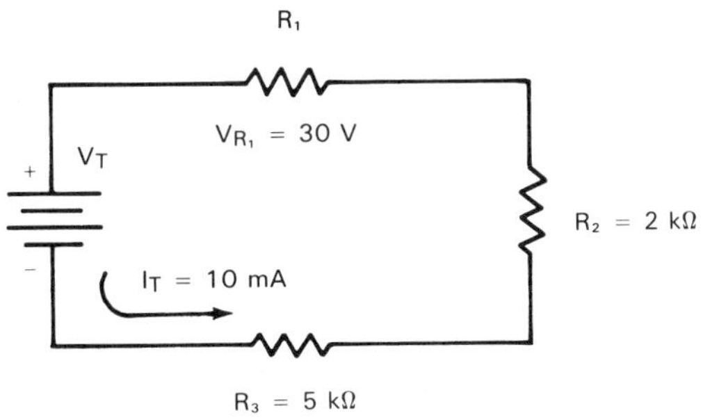

a. Value of resistor R_1 13a. ____________

b. Total resistance (R_T) b. ____________

c. Voltate drop across R_2 (V_{R_2}) c. ____________

d. Power dissipated in R_3 (P_{R_3}) d. ____________

14.

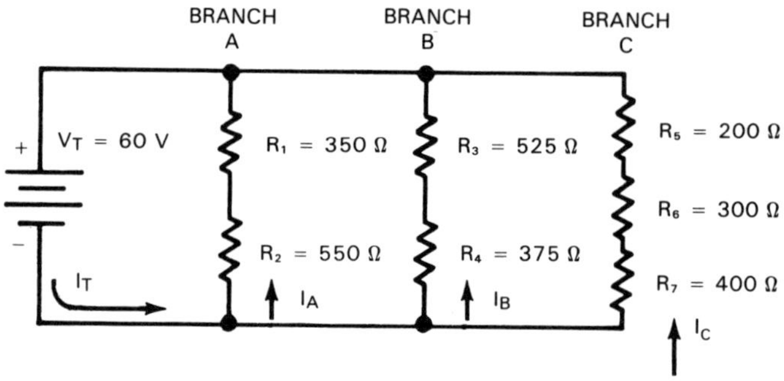

a. Total resistance (R_T) 14a. ________________

b. Total current (I_T) b. ________________

c. Current in Branch A (I_A) c. ________________

d. Current in Branch B (I_B) d. ________________

15.

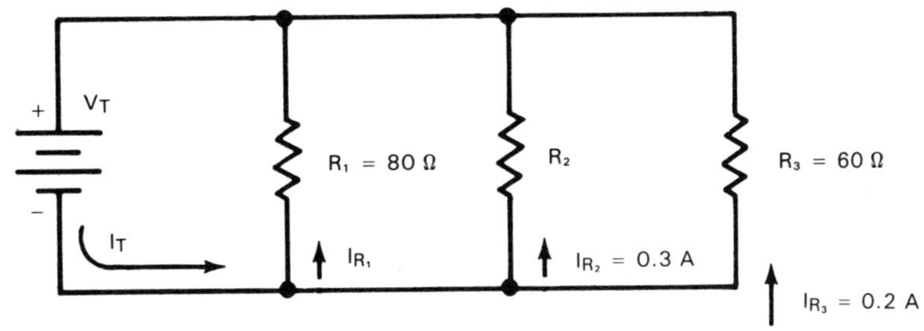

a. Value of resistor R_2 15a. ____________________

b. Total resistance (R_T) b. ____________________

c. Branch current I_{R_1} c. ____________________

d. Applied circuit voltage (V_T) d. ____________________

16.

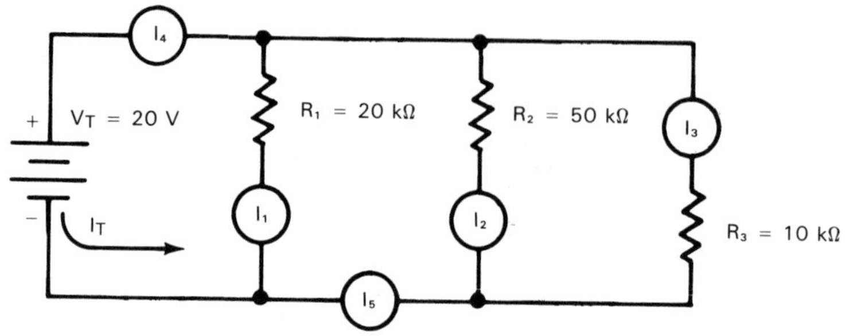

a. Current in meter I_1 16a. ______________

b. Current in meter I_2 b. ______________

c. Current in meter I_3 c. ______________

d. Current in meter I_4 d. ______________

e. Current in meter I_5 e. ______________

Chapter 18

INTRODUCTION TO ESTIMATING

Test Score ______________

Name ______________________________

Date ______________ Class ______________

Complete each of the following problems. Show your work and place answers on the lines provided. Include units where necessary.

1. Convert the linear measurements to board feet.

Nominal	Linear Length	Board Feet
a. 2 × 4	100 inches	________
b. 2 × 6	60 feet	________
c. 2 × 10	4 feet	________
d. 1 × 8	580 inches	________
e. 1 × 12	48 feet	________

2. Convert the board-foot measurements to linear feet.

Nominal	Board Feet	Linear Length
a. 2 × 4	24 bd. ft.	________ inches
b. 2 × 6	96 bd. ft.	________ feet
c. 2 × 10	328 bd. ft.	________ inches
d. 1 × 8	150 bd. ft.	________ feet
e. 1 × 12	36 bd. ft.	________ feet

3. a. Determine the exact length of pressure-treated, 2 × 6 nominal (1 1/2″ × 5 1/2″ actual) lumber needed as a sill for the concrete foundation of a house measuring 46 feet wide by 60 feet long.

3a. ____________________

b. How many 12-foot long 2 × 6s will be required for the sill?

b. ____________________

4. Determine the total length of 2 × 4s needed for a double top plate and single bottom plate to construct a wall 46 feet long. 4.________________

5. What is the total length of 5 1/2 inch wide board needed to trim around a 3-foot by 4-foot window? 5.________________

6. Calculate the number of framing members needed:

Stud Spacing	Wall Length	Number
a. 16″ O.C.	16′	________
b. 16″ O.C.	21′	________
c. 16″ O.C.	315″	________
d. 24″ O.C.	24′	________
e. 24″ O.C.	120″	________

Answer questions and round to the nearest tenth, if necessary.

7. How many 4-foot by 8-foot sheets of plywood are needed for the subfloor of a 32- by 36-foot house? 7.________________

8. Determine how many 2-foot by 4-foot tiles are needed to cover a ceiling in a 16- by 20-foot room.

8.____________________

9. Calculate the cubic yards of concrete needed for an 8-inch-thick concrete wall that is 6 feet high and 32 feet long.

9.____________________

10. Calculate the cubic yards of concrete needed for a 4-inch-thick concrete floor that is 24 feet wide and 36 feet long.

10.____________________